CONTENTS

46

16

ON THE COVER:
Developed by maker Rongzhong Li, NybbleCat
is an open source quadruped robot kit that
walks, waves, reacts to commands, and looks
super cute doing it.

Artwork: Rongzhong Li

PROJECTS

106

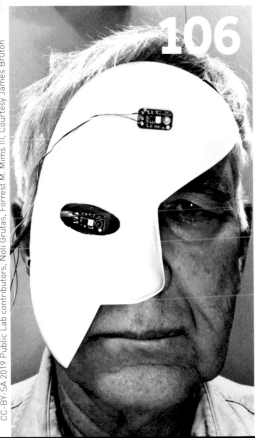

118

SKILL BUILDER

TOOLBOX

SHOW & TELL

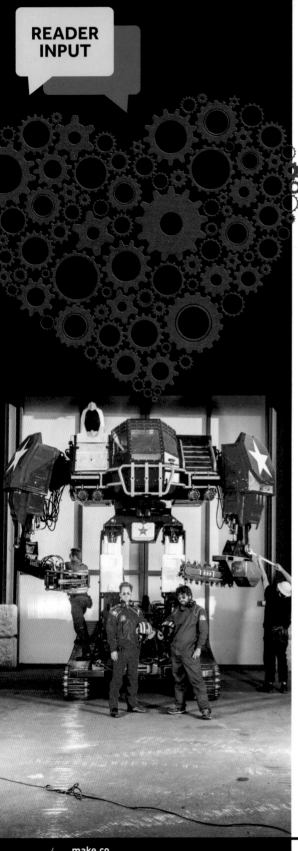

FEELIN' THE LOVE

FINDING A PATH

One day in sixth grade I was in my school library playing with some Snap Circuits with some other kids when it was time to clean up. I stood over by the magazines and picked one up to act like I was doing something. I glanced at the magazine page. It had a picture of the MegaBots team and an article about it. As soon as I saw this I sat down and read the magazine cover to cover. With every page I saw I began to realize more and more that this is where I belong. I was born to create and contribute to my community by building things.

I read all that I could on your website and I made a little workshop in my garage. I would spend hours and hours at that little workbench making what ever came to mind. I taught myself as many skills as possible, from soldering to 3D printing to coding. Every time I would sit down at that tiny workbench in my parent's garage my mind would run wild at the endless possibilities before me. And all the while I was reading every copy of *Make:* magazine under the sun and dreaming of maybe even working there someday. And now as I write you this email, I can see a path before me, a path to NASA or MIT or even Boston Dynamics, a world that I can shape and build to my imagination.

You will never know how many lives you have touched across the globe. Kids like me and adults too, you gave us all a place to call home and a path to the stars. Who knows what I will create as I grow, maybe a rocket to Mars or lifesaving machines. What ever it is, just remember, it is all because of you. All because, I picked up your magazine, on a rainy day, and found out who I am meant to be. — *Sam Millar, age 14*

 vol. 02

Make:
technology on your time

Make an HDTV Recorder Before It's Too Late

29 HOME ENTERTAINMENT PROJECTS

Build a Light-Seeking Robot from an Old Mouse

+ PODCASTING **101**

Retro-gaming Hacks

Home Theater Shaker Seats

Instant Surround Sound

R2-DIY EXTREME BOT BUILDERS AT HOME

O'REILLY makezine.com

A FAMILY AFFAIR(E)

I came across an email many years ago when *Make:* Vol. 02 had just come out. I quickly purchased and snatched up a copy of Vol. 01, subscribed and have looked forward to each copy since then. I still remember the project with the kite photography using Popsicle sticks to trigger the camera as one of my favorite projects.

My oldest son Josh was 12, and my younger Levi was 7 when the first Maker Faire was held, and we had never done a road trip together and really only done the standard cub scout type trips as a "guy thing" as my wife says. We had no idea what to expect (the same as everyone else the first year!). I got us a hotel reservation, and bought some tickets and we climbed in the Mini Cooper and made the 10–12 hour trip (depending on how many bathroom stops we made) from Oceanside up to San Mateo on a Friday afternoon/evening. We got up the next morning and as we pulled into the San Mateo fairgrounds parking lot we spied huge plumes of flame jetting 30–40 feet in the air... We knew we had found the right place!

We have not missed a single Maker Faire since and Josh is now married and graduated college with a degree in mechanical engineering. Levi is going into his second year of college and is working toward a degree in mechanical engineering technology. Along the way we have brought a dozen different people to the Faires and we have a fairly consistent crew of 8-10 that make the trek every year in our pilgrimage up to San Mateo. — *Eric Warner, via email*

Come Together

 PowellBears
@PowellBears

Yesterday we got married @makerfaire , where we met exactly 7 years ago to the day. Congratulated by @make photographers and @makerfaire execs and fellow attendees. Sorry, boss @simsea - can't come in today - calling in JUST MARRIED.

Make: @make · May 19
Surprise wedding at @makerfaire! #MFBA19

9:22 AM · May 20, 2019 · Twitter for Android

 Dorothy Jones-Davis
@dmjonesdavis

When I go to #mfba19, it's like coming home. No matter how many times I've been, I can always find something new & intriguing - projects, panels, & makers who are all amazing. So honored to be on the #mfba19 Advisory Board & so happy to support such a great celebration of making!

5:55 PM · May 18, 2019 · Twitter for iPhone

Evan & Katelyn
@EvanAndKatelyn

The best part of events like @makerfaire are the awesome people and community! ♥ #MFBA19

👤 Joel Telling - 3D Printing Nerd and 8 others

10:35 AM · May 18, 2019 · Twitter for Android

Make:®

"If you're alive, you're a creative person. You and I and everyone you know are descended from tens of thousands of years of makers." – Elizabeth Gilbert

PRESIDENT
Dale Dougherty
dale@make.co

VP, PARTNERSHIPS
Todd Sotkiewicz
todd@make.co

EDITORIAL

EXECUTIVE EDITOR
Mike Senese
mike@make.co

SENIOR EDITORS
Keith Hammond
keith@make.co

Caleb Kraft
caleb@make.co

PRODUCTION MANAGER
Craig Couden

CONTRIBUTING WRITERS

Anthony Altorenna, Chris Anderson, Nathaniel Bell, Axel Borg, Sam Brown, Adam Conway, Bronwen Densmore, Shannon Dosemagen, Marcus Dunn, Toni Klopfenstein, Sonny Lacey, Britt Michelsen, Forrest M. Mims III, Camilo Parra Palacio, Alex Reveles, Rongzhong Li, Josh Tulberg, Alexey Volochenko, Robert van de Walle

DESIGN & PHOTOGRAPHY

CREATIVE DIRECTOR
Juliann Brown

MAKEZINE.COM

ENGINEERING MANAGER
Alicia Williams

WEB APPLICATION DEVELOPER
Rio Roth-Barreiro

GLOBAL MAKER FAIRE

MANAGING DIRECTOR, GLOBAL MAKER FAIRE
Katie D. Kunde

MAKER RELATIONS
Sianna Alcorn

GLOBAL LICENSING
Jennifer Blakeslee

MARKETING

DIRECTOR OF MARKETING
Gillian Mutti

OPERATIONS

OPERATIONS DIRECTOR
Cathy Shanahan

ACCOUNTING MANAGER
Kelly Marshall

OPERATIONS MANAGER & MAKER SHED
Rob Bullington

PUBLISHED BY

MAKE COMMUNITY, LLC
Dale Dougherty

Copyright © 2019
Make Community, LLC. All rights reserved. Reproduction without permission is prohibited.
Printed in the USA by Schumann Printers, Inc.

Comments may be sent to:
editor@makezine.com

Visit us online:
make.co

Follow us:
🐦 @make @makerfaire @makershed
📘 makemagazine
📷 makemagazine
▶ makemagazine
📺 twitch.tv/make
📌 makemagazine

Manage your account online, including change of address:
makezine.com/account
866-289-8847 toll-free
in U.S. and Canada
818-487-2037,
5 a.m.–5 p.m., PST
cs@readerservices.
makezine.com

Make:
Community

Support for the publication of Make: magazine is made possible in part by the members of Make: Community. Join us at make.co.

CONTRIBUTORS

What's an outdoor project you hope to build someday?

Britt Michelsen
Mölln, Germany
(Soap Vomiting Unicorn!)

Two things: a motorized unicorn float and a 3D printed Squirtle (Pokémon) fountain.

Nathaniel Bell
Hillsbourough, North Carolina
(Make a Sweet Aluminum Whistle)

I've always wanted to make a small boat, like a canoe or kayak.

Camilo Parra Palacio
Olomouc, Czech Republic
(Makey Is Real!)

I would like to build an automated smart gardening system for our garden.

Issue No. 70, October/November 2019. *Make:* (ISSN 1556-2336) is published quarterly by Make Community, LLC, in the months of January, April, July, and October. Make Community is located at 708 Gravenstein Hwy N, Box #239, Sebastopol, CA 95472. SUBSCRIPTIONS: Send all subscription requests to *Make:*, P.O. Box 17046, North Hollywood, CA 91615-9588 or subscribe online at makezine.com/offer or via phone at (866) 289-8847 (U.S. and Canada); all other countries call (818) 487-2037. Subscriptions are available for $34.99 for 1 year (4 issues) in the United States; in Canada: $39.99 USD; all other countries: $50.09 USD. Periodicals Postage Paid at San Francisco, CA, and at additional mailing offices. POSTMASTER: Send address changes to *Make:*, P.O. Box 17046, North Hollywood, CA 91615-9588. Canada Post Publications Mail Agreement Number 41129568. CANADA POSTMASTER: Send address changes to Make Community, PO Box 456, Niagara Falls, ON L2E 6V2

Shape Our Future Together

by Dale Dougherty, Founder and President of Make: Community

It's good fortune to have others pick you up when you're down. I got this email in June from Bryson Turner, a 14-year-old maker from Chicago.

I read an article about you shutting down Maker Media due to funds and support. However, it also said to write to you to tell you what impact you had on us makers. It was amazing when I attended the Bay Area Maker Faire last year and even more amazing when I went and became a maker this year. Maker Faire means so much to me that my family and I are willing to fly out to California for it from Chicago. Maker Faire was an awesome experience, and I will always cherish the memories made there.

Make: magazine was awesome as well. It has always helped me come up with new ideas. I have been a subscriber for over three years now.

Overall, the maker society has shaped my life through influence, empowerment, and inspiration. "Maker" has become a common household word with my family. It may be a simple word, but the meaning behind it is truly great. Not only does maker mean inspiration, it means the innovations and a path towards the future. Maker is a new future, a place where everyone who wants to create can, who wants to learn can, or who wants to try new opportunities can. Maker Faire is a space where people of all different religions, races, and ages can come together and speak freely. This is where people/innovators can connect, contrary to people telling you the ideas you have are bad. As the honorable Adam Savage said during his Sunday sermon during the 2019 Bay Area Maker Faire, "We come together as a family; where other people will not understand us, we understand each other. We come together to shape our future together." These words have great power, and were shaped because of the maker company.

I believe in Maker Faire. It is a great idea. If there is any way I can help, that would be awesome. Mr. Dougherty, I believe in your company and its idea. So far, you have been doing great. Please allow me to help because I believe in the maker community.

From,
An innovator
A believer
A maker

Thank you, Bryson. I truly appreciate hearing from you and many others who have told me about the impact of Make: and Maker Faire on their lives.

Madelin Wood wrote me:
"There has always been such a strong sense of belonging within the Make community. The culture you've inspired, and people it's attracted has touched people's lives in ways you may never know. I sincerely appreciate every opportunity you've provided for me personally, and how much you've made possible for every kid and grown-up with a dream and an itch to build."

That's why we will keep going, not just publishing this magazine but supporting the community that has grown up around both tech-enabled and traditional practices of making.

With Make: Community at make.co, this is more than just a reboot. It represents a fundamental shift in focus as we become a member organization, an association of makers whose goal is to serve the community and grow it. Please consider becoming a member. Together, we can and should shape the future by empowering and connecting more makers. Our collective intelligence and creativity can positively change the world in small and large ways. We already have.

I'd like to hear your story and how you can help shape this future together. Write me at dale@make.co. ◗

Join Make: Community at make.co.

MADE ON EARTH

Backyard builds from around the globe

Know a project that would be perfect for Made on Earth?
Let us know: *editor@makezine.com*

ANATOMICAL ARTISTRY

MICHAELALM.COM

What do you do on your lunch break? Artist **Michael Alm** (Instagram: @michael.alm; Youtube: michaelalm; Facebook: almsculpture) used to take his break from the art museum at the University of Washington campus and wander through the nearby natural history museum. During these lunch outings, while relaxing and drawing the fossils and taxidermied animals on display, he began to see a similarity between the bunched muscle on the animals and the stacks of thin wooden strips being tossed aside as a waste byproduct of an art installation at work.

He began hoarding the strips, lugging them to his home studio for experimentation. Drawing from his experience as a furniture maker, he knew that he could wet the strips and form them, gluing them in layers to create three-dimensional shapes.

Aside from the eyes, his sculptures are made entirely out of wood. There are thin strips of veneer as muscle, hand-sculpted pieces for bones and texture, and wood shavings as fur or feathers. The results are simply stunning in their ability to convey life and motion.

Though it may be tempting to look at these as pure anatomical displays, they are artistic interpretations of the animal. Alm considers the placement of each sinew and muscle.

"The gaps in the veneer accentuate the tension in the form while lightening the visual weight of the creature," Alm says. "In *Jackrabbit (Lepus californicus)*, I've highlighted the elements [that] contribute directly to the animal's movement and eliminated any excess. As a result, the form looks both strong and delicate much like the animal itself."

Alm's art has been featured in a plethora of publications and museums and he continues to explore new ways of celebrating the fauna that inspire him. —*Caleb Kraft*

Michael Alm

"LITTLE GYPSY" LOWRIDER TIJUANARICK.COM

Lowriders are truly an impressive form of art. Crafting their rich and glossy paint jobs takes an incredible amount of time and effort, as does the engineering that goes into making the suspension do exactly what you want, including lifting and dropping the car in ways it was never initially intended to do.

On a great build, effort simply isn't enough; you need a muse. For some, this is their heritage. For others, it is inspiration from pop culture. For **Ricardo Cortez**, aka **Tijuana Rick**, his muse was his daughter. Like the age-old tale of a parent crafting a crib for their child, Cortez fabricated a custom miniature 1962 Impala for his little girl.

"I've been creating for a long time, pulling inspiration from my San Jose car culture and personal upbringing," Cortez says. "When my wife and I found out we were going to have a girl, I began to brainstorm endless possibilities of projects, finally settling on something that hadn't been created before — a 1962 R/C lowrider pedal car. This not only would provide joy for her as we would pique the interest of spectators."

This isn't your standard pedal car; Cortez sunk his heart and soul into the build. It is radio controlled and has a custom frame and suspension that allow it to raise and lower just like the cars at the shows he brings her to. He gave it a beautiful paint job and topped it off with a plushy interior to cradle his daughter in safety and comfort.

Cortez had to stretch his skills a bit with this build. He wanted custom wheels and decided to design and 3D print them on his own, meaning he had to learn an entirely new toolset.

"I had never 3D modeled before and taught myself Fusion 360. Through trial and error, I was finally able to print an acceptable tire and wheel combination true to the 1970s-style wheels of original lowriders."

Though his daughter may be too young to fully appreciate the skill and love Cortez put into this project, it is surely going to be a hit at car shows, where she'll be collecting memories that will last a lifetime. —*Caleb Kraft*

© Ricardo Cortez

Justin Gray's
*Cadillac Rocket
Car MK3* came
alive with jets
of fire for Faire-
goers throughout
the weekend.

A FAIRE TO REMEMBER

Despite the rainy weather and an uncertain future, Maker Faire Bay Area 2019 was a blast!

Maker Faire Bay Area had a bittersweet undertone this year. But even with a steady drizzle throughout the weekend, the large crowds were still excited by the myriad sights to behold, and many deemed it their favorite event in years. Here are just a few of the projects on display. For even more pictures and videos, check out

Long time supporters and TV personalities **Conan O'Brien** and **Adam Savage** connect.

Hiroaki Suzuki is a Japanese hardware engineer who danced around the Faire wearing his **life-sized, rod puppet skeleton**.

Adam Savage made his way to his "Sunday Sermon" stage in **Justin Gray's** *Capture Robot Squid*.

Bay Area R2 Builders' life-sized, fully functional **Star Wars** droids are always a crowd favorite.

Constructed of repurposed materials, wheel chair components, and over 1,300 addressable LEDs, **David Date's** *Myriapoda* is a 30-foot mechanical centipede that lit up this year's dark room.

Becca Henry, Chris Willis

MADE ON EARTH

Adam delivered his "Sunday Sermon" from **Bruce Tomb's** electric *Maria Del Camino*.

The Museum of Future Sports featured free hands-on experiences with **drones, RC racing,** and **virtual reality** to Faire attendees.

LudoSport is a **lightsaber combat sport** that trains athletes to safely compete in tournaments.

A Maker Faire staple, **Acme Muffineering** builds electric vehicles in the shape of **cupcakes** that zip around at up to 15mph.

The **giant Lite Brite** in Expo Hall was easily a crowd pleaser.

Becca Henry

EPC-M.ZONE, a group of makers from Taiwan and Singapore, brought **incredible cardboard creations** with lights and audience-activated sound effects.

Shawn Thorsson (author of *Make: Props and Costume Armor*) showed off a host of sci-fi and fantasy-themed props, including these **Warhammer 40,000 outfits**.

Ryon Gesink's *F1 Machine* is a rolling electric desert vehicle that spins on massive wheels and spits fire.

Regardless of whether the San Mateo Event Center is overrun next year with fire-spewing art pieces, hands-on workshops, or robot critters underfoot, Maker Faire as a concept is still going strong. More than 200 events are taking place around the globe. Find a Faire near you on page 105 or at makerfaire.com/map. 🅐

WRITTEN BY SHANNON DOSEMAGEN

COMMUNITY SCIENCE

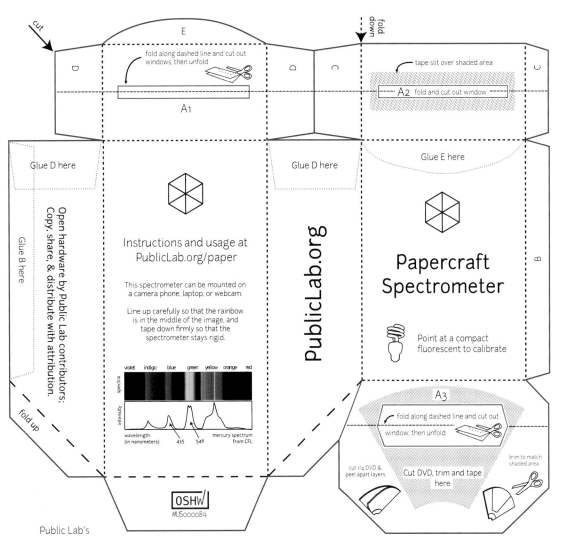

cut

fold down

E

D

fold along dashed line and cut out windows, then unfold

A1

D

C

tape slit over shaded area

A2 fold and cut out window

C

Glue D here

Glue D here

Glue E here

Open hardware by Public Lab contributors; Copy, share, & distribute with attribution.

Glue B here

Instructions and usage at PublicLab.org/paper

This spectrometer can be mounted on a camera phone, laptop, or webcam.

Line up carefully so that the rainbow is in the middle of the image, and tape down firmly so that the spectrometer stays rigid.

PublicLab.org

Papercraft Spectrometer

Point at a compact fluorescent to calibrate

B

violet indigo blue green yellow orange red

spectra

intensity

wavelength (in nanometers)   435   546   mercury spectrum from CFL

fold up

A3

fold along dashed line and cut out window, then unfold

trim to match shaded area

cut 1/4 DVD & peel apart layers

Cut DVD, trim and tape here

OSHW
#US000089

Public Lab's

Papercraft Spectrometer v2.0.9

Open Source CERN OHL v1.2

This open source hardware design was developed by contributors like you. Have questions? Ideas for improvement? Want to collaborate with others?
Visit: PublicLab.org/paper

This template can be photocopied for reuse! For best results, use cardstock or a similar heavy paper. Make sure to do double-sided copies so the black box and slit cutout will print on the opposite side.

What you'll need:
- scissors
- tape or glue
- a butter knife or ballpoint pen
- a DVD-R

Instructions
more info at PublicLab.org/paper

1. Cut out the spectrometer along the red lines (these may be grey if photocopied in black & white)

2. Use a butter knife or ballpoint pen to score the dotted lines for folding

Fold over on dotted lines, then cut out window

3. Fold over and cut out the "windows" at A1, A2, & A3

Mountain fold   Valley fold

4. "Mountain" fold along the dotted lines: ··········

5. "Valley" fold along dashed lines: ― ― ―

6. Cut, peel, and attach DVD-R fragment (see A3 for directions)

slit

tape

7. Assemble the slit cutout on the opposite side. Tape the slit over window A2 and trim as needed.

8. Align tabs B, C, D & E and tape or glue them down.

9. Visit PublicLab.org/paper to learn how to calibrate and use your new spectrometer!

Instructions: PublicLab.org/paper
Open Hardware - CERN OHL 1.2

1. Cut out bar and the U-shape below it,
   keeping lines as straight as possible
2. Move upper piece down to form
   the slit.
3. Align strip above the dotted line,
   leaving a narrow gap for light.
   Glue or tape in place.

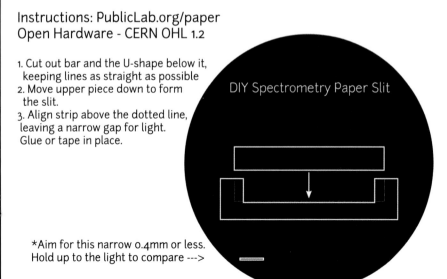

DIY Spectrometry Paper Slit

*Aim for this narrow 0.4mm or less.
Hold up to the light to compare --->

**SHANNON
DOSEMAGEN**
is the executive
director of the Public
Lab nonprofit. With
almost two decades
of experience in
community organizing
and education,
she has worked
with environment
and public health
groups addressing
declining freshwater
resources and coastal
land loss, and on
building participatory
monitoring programs
with communities
neighboring industrial
facilities and impacted
by the BP oil spill. She
lives in New Orleans,
Louisiana
Twitter: @sdosemagen

In May 2010

I set out early one morning with a small group of concerned Gulf Coast residents, headed to meet a boat captain who would take us to the Chandeleur Islands, a chain of barrier islands in southeast Louisiana. Within minutes of leaving the harbor and navigating out into the Gulf of Mexico, the wind picked up, sending water sloshing back and forth across the boat, soaking us all to the bone. We went by shrimp trawlers eerily turned into skimming vessels. We passed the remnants of barrier islands rendered almost non-existent during Hurricane Katrina in 2005. And we navigated past seemingly post-apocalyptic clusters of oil pipelines jutting out of the water like a scene out of *Waterworld*. This was in the first weeks after the BP oil disaster — what would become the largest oil spill in U.S. history, leaking an estimated 3.19 million barrels and impacting over 1,000 miles of coastline.

Once we made it to the Chandeleur Islands, our boat captain began surveying the damage to his fishing grounds. He cut the engine and we readied our equipment: a bag filled with rubber bands and Q-tips, the top half of a two-liter soda bottle converted into an aerial stabilizing rig, a large balloon, kite string and reel, a digital camera, and duct tape (always duct tape!). We launched the rig into the air and watched as the big, white balloon became indistinguishable from the bright sun. I remember the swirl of rainbows in the water and the nauseating wooziness from a heady mixture of dispersant chemicals, petroleum fumes, and the heat of a Louisiana day.

We were organizing as a group called "Grassroots Mapping." With our DIY balloon rig and camera, we collected the first few thousand aerial images of the spill. In the weeks that followed, we would gather tens of thousands more images, collected by hundreds of concerned Gulf Coast residents — people coming from Texas, the Florida Panhandle, and everywhere in between to document the extent of the damage to the beaches and wetlands, fisheries and wildlife. We knew it was important to share a different narrative than the one that was being shared by BP. Photos and the resulting maps we created with our "community satellites" showed the damage and the scale of the disaster, and would go on to be shared with media around the world.

Mapping invasive species on
New Orleans' Bayou Bienvenue

Chandeleur spill damage documented by Public Lab's DIY balloon rig.

In the wake of the spill, we began to question assumptions about how science could be done in public spaces; we wanted to show that *all* people could bring valuable input to asking and answering environmental health questions. Citizen science had long been a growing field of practice, but at that time, it was largely driven by scientists who used data collected by people in local communities. We sought to redistribute power by making science something that *anyone* could access — where people with different forms of expertise are recognized for the value of lived experience and local know-how. We envisioned a network in which people down the block or on the other side of the world could work together on solutions, learning from each other's experiences and problem-solving. It was through these ideals that Public Lab and our movement were born.

We've come to call our work "community science" — a science that supports the organizing

Mapping invasive water chestnuts on Lake Warner, MA

activities and goals of everyday people who are working to answer their own environmental questions. Organizing models from groups such as the regional Bucket Brigades, who popularized a type of low-cost community air sampling, taught us the importance of collecting data in places with deep social meaning — on the steps of churches or on the local elementary school playground. We also knew we needed to disrupt the paradigm of technology created solely for industry, government, and research institutions, instead employing an alternative model to proprietary tools that "black boxed" processes and created hardware that couldn't be modified or improved upon. Public Lab operates under a strong commitment to open source hardware principles: relying on collective input from people toward the development of tools, technology, and methodology, coupled with open hardware licensing, to ensure that anyone in the world can continue to develop and build upon past work.

Nearly a decade since the BP oil disaster and the founding of Public Lab, we're inspired to see how community science, making, and open hardware are taking hold around the world. In Ghana, an organization called Global Lab is focused on bridging science with public life through knowledge exchanges, mentorship, community action in rural areas, and discussion around topics such as water sanitation. In a recent project, they partnered with OpenAQ to release a community statement on air quality in Ghana, making a case for open data to be part of a solution towards people understanding air pollution and the effects. Building on this work and a datathon that produced maps, predictive models, and prototypes around air inequality in Accra, they are interested in exploring longer term how open hardware can support the development of monitors to measure air pollution and work with residents on understanding the drivers and causes of poor air quality. And projects with similar goals as Public Lab are taking root all over, from mobile and fixed particulate monitoring in urban areas, to water monitoring in the Amazon, accessible open source hardware movements around microscopy and spectrometry, to exploring marine environments with DIY microplastic trawlers.

Mapping replanting efforts on the Yellow Bar Hassock with American Littoral Society and Dredge Collaborative

WE CALL OUR WORK "COMMUNITY SCIENCE"

Navigating the Gowanus Canal with Gowanus Low Altitude Mapping (GLAM)

Inherent in all is a people-centered science and the drive to create and make tools that will help us all understand and respond to the environmental and health questions we come in contact with on a daily basis.

With partner groups and people around the globe, Public Lab is working to build a healthier and more equitable world through community science, open technology, and environmental justice — supporting a new generation of people equipped with the means to ask questions, collect and interpret data, and advocate on behalf of their own communities. It's both a critical and exciting time for scientists and the maker community to get involved in helping to address the environmental issues of our times. We encourage you to learn from and find inspiration in the stories presented in this section and hope that they will provide a starting point for your interest in making towards meaningful impact. ◗

Learn more at publiclab.org/make and follow our social channels for live builds of the projects here: @PublicLab

BUILD YOURSELF SOME SCIENCE:

Inside **Public Lab's** Open Source Tools **WRITTEN BY BRONWEN DENSMORE**

Monitoring stormwater with youth from Groundwork New Orleans

IN the summer of 2012, I came across an event listing for a kite building workshop and the next thing I knew I was on a loading dock in Brooklyn, NY, building kites and solar balloons with two of Public Lab's founders. Seven years later, I am Public Lab's Open Hardware Community Manager — one of the people who reaches out to people who might be interested in building, learning or exploring with DIY environmental tools.

Public Lab members are people who are passionate about finding ways to build and use tools to solve real-world research problems, and who use those problem solving skills to connect with and improve aspects of their communities. We're excited about the contributions that everyone brings to the table: part makerspace, part research community, part organizing platform, Public Lab brings people who are interested in making change together.

Getting involved with Open Hardware projects is as easy as showing up (in person at an event, or joining an online site or forum), introducing yourself and expressing an interest. Public Lab celebrates a diversity of projects, skills, and types of contributions — in addition to tech oriented projects, our Open Hardware contributors include gardeners, emergency responders, illustrators, kite and balloon enthusiasts (obviously!), biologists, community organizers: people with skills of every type and level who are excited to learn and share with others. We run regional events called Barnraisings designed to bring local groups, environmental organizers, and DIY folks together. We present our work at Maker Faires, and hold workshops around the world so that other people can connect with the things that a big, open community has to offer.

BRONWEN DENSMORE is Public Lab's Open Hardware Community Manager. She comes from background in art, fabrication and academic librarianship. She lives in Brooklyn, New York.

Assembling kits at the Texas Barnraising event

Collaborating on satellite receiving hardware at the Cocodrie, LA Barnraising event

Coqui conductivity meter

PUBLIC LAB BRINGS PEOPLE WHO ARE INTERESTED IN MAKING CHANGE TOGETHER.

We welcome anyone with an interest in using or making tools in their local environment to reach out to us — we're excited about living in a world where applying our skills (maker skills or otherwise) to environmental issues can lead to a healthier world for all of us.

Here are a few of the kits from our community.

COQUI

The Coqui (named for a Puerto Rican frog with an especially loud and recognizable call) determines conductivity in water. This tool doesn't require a lot of equipment or electronics experience to build: the circuit measures conductivity between two wires, and communicates that information through a speaker as a that tone increases in pitch as conductivity increases. In something like distilled water, this will be relatively low, but with the introduction of materials with greater conductivity

(salt is one example), it will increase.

High or low conductivity isn't necessarily bad — sea water will naturally have a higher conductivity than fresh water, for instance, but using a tool like the Coqui to test multiple spots along a waterfront to scan for trends or changes in the conductivity can help you to identify spots where something might be entering or moving through the environment—either from the built environment (say, runoff from an industrial site, or an underwater drainage pipe), or naturally, as fresh water enters the system.

This tool was created as part of the Open Water Project at Public Lab, designed in part for use in education. You can use this along with a short handbook developed by Catherine D'Ignazio called *Sensor Journalism: A Guide for Educators*, which outlines ways that data collection can be used to tell compelling stories about issues in the environment.

COQUI INSTRUCTIONS:
publiclab.org/coqui

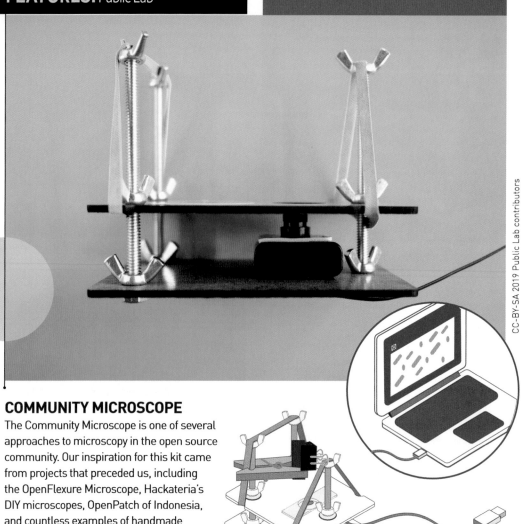

COMMUNITY MICROSCOPE

The Community Microscope is one of several approaches to microscopy in the open source community. Our inspiration for this kit came from projects that preceded us, including the OpenFlexure Microscope, Hackateria's DIY microscopes, OpenPatch of Indonesia, and countless examples of handmade magnification tools that people have built and used for centuries. This version was developed in collaboration with Parts & Crafts (a makerspace and community workshop in Sommerville, MA).

Public Lab's microscope can be assembled using a variety of materials and techniques — two platforms hold a camera and a slide, and a simple mechanism made from rubber bands and wingnuts allows a user to focus the device. The simplest version of the microscope uses a USB webcam with a flipped lens, but other versions can be made with higher quality lenses or traditional microscope objectives. Raspberry Pi cameras can let you connect over Wi-Fi, and changing the orientation of the stages and lights let you experiment with different ways to view a sample.

This kit is especially great for examining microorganisms in water samples. The basic webcam it has a field of view of about a millimeter — small enough to clearly see things like plankton, amoebas, and tardigrades! It can also be paired with a sample collection tool like BabyLegs (page 24) as you investigate different kinds of pollution in your water.

MICROSCOPE INSTRUCTIONS:
publiclab.org/micro

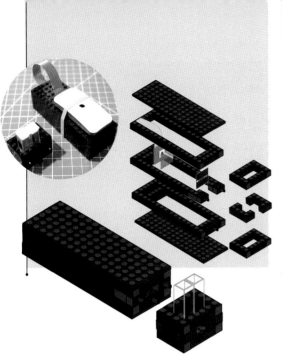

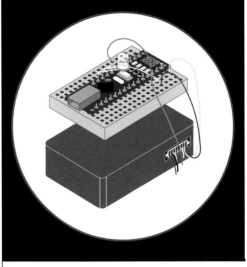

LEGO SPECTROMETER

This tool helps analyze the chemical composition of a sample. By passing light through a sample and scattering it so that it is captured by a camera as bands of color, and then transforming those bands into a graph that gives more precise data about highs and lows, it is possible to start identifying chemicals that may be present in the sample.

Public Lab users generally share their results on Spectral Workbench (spectralworkbench.org), an online tool that will convert your images (they'll look like rainbows) into graphs. The highs and lows on various points of the graph can help you to identify chemical components in a sample, and by comparing the spectra of an unknown sample to the spectra of something that has already been identified it is possible to draw conclusions about what may (or may not) be present.

The spectrometer and Spectral Workbench are some of Public Lab's earliest tools; by breaking an otherwise expensive tool into more accessible/DIY parts (Lego or cardboard, webcams or Raspberry Pi cameras, recycled DVDs) it means that spectroscopy is a research method that anyone can explore.

SIMPLE AIR SENSOR

The Simple Air Sensor is a prototyping and onboarding kit to help you get started learning the basics of particulate matter air pollution, electronics, and the design of experiments, without having to learn how to program. It's inspired by open source work from Julieta Arancio, Gustavo Olivares, and many others.

On the unit, a realtime readout changes color when the sensor reads differences in air quality. You can study the performance of the sensor itself by watching changes in the colored light readout in the presence of different sizes of particle (dust) in different humidity conditions.

IMPORTANT: Be aware that this red/**yellow**/**green** light only illustrates the readings of PM2.5 levels and is not converting them into an Air Quality Index (AQI) that indicates health risks.

This sensor circuit uses the same type of particle detector as the PurpleAir sensor, but differs (and is cheaper) as it only uses one as compared to PurpleAir's two. It also does not include a temperature/humidity/pressure sensor as the PurpleAir does, and instead of logging into either the cloud or to the device itself for readings, ours displays its changes via the colored light.

LEGO SPECTROMETER INSTRUCTIONS: publiclab.org/lego

SIMPLE AIR SENSOR INSTRUCTIONS: publiclab.org/simple

BABYLEGS

This simple DIY trawl can be used to collect water samples for analysis. A trawl works by letting water pass through a net while materials are caught inside. Usually a trawl is towed behind a boat or set up along a shoreline, river, or someplace where running water can pass through it. What remains in the net (in this case, made from a pair of pink tights!) can be transferred to a sample bottle and examined under a microscope.

BabyLegs was developed by Max Liboiron and the Civic Laboratory for Environmental Action Research (CLEAR) to collect samples for marine microplastics research (see page 27). Microplastics enter our waste stream as larger plastic breaks down and become an issue as they are consumed by marine animals and absorbed into the environment, putting people, animals and marine ecosystems at risk. Samples gathered from the ocean surface can be examined under a microscope. Even though microplastics are quite small, in many cases it is possible to identify the type of plastic and hypothesize its source — this not only helps us to recognize what issues we may be coping with locally, but also track how waste travels through marine environments.

Trawls like BabyLegs can also be used to investigate the microbiology of an ocean, pond, lake or creek (really, any body of water). Microorganisms are also captured in trawl nets — organic matter (sediment, plant matter) can be home to all kinds of tiny organisms that you can see under a microscope.

BABYLEGS INSTRUCTIONS:
publiclab.org/babylegs

INFRAGRAM PI

The Infragram Pi camera was designed to help visualize plant health. As plants convert light energy into chemical energy via photosynthesis, most plants absorb light from the blue and red range of the spectrum and reflect away the infrared range. By using cameras that can capture infrared, it is possible to get a sense of how well a plant (or group of plants) is converting light into food. By comparing pictures of similar plants growing in different conditions, it is also possible to hypothesize about the environmental conditions that might be affecting plants. One example might be in a controlled experiment, where two identical plants are being grown in different kinds of soil or fertilizer. In environmental work, people use infrared (IR) photography to survey larger areas in the hopes of seeing how and where soil and vegetation may be impacted by pollution. An easy way to survey a large space (like a field, or waterfront) is to capture photographs from above — lots of Public Lab contributors use kites or balloons to do aerial mapping.

The Infragram Pi camera is small and lightweight, which makes it a good choice for aerial photography. It uses either an infrared Raspberry Pi camera (these can be purchased), or an existing camera with its infrared filter removed. Infragram software (available through Public Lab's store and website) lets a user connect to a live camera feed over a Wi-Fi network and download to a memory card. Once images are downloaded, the colors can be processed on infragram.org or Image Sequencer (an image processing tool developed with the support of the NASA AREN project).

INFRAGRAM PI INSTRUCTIONS:
publiclab.org/infragrampi

PUBLIC LAB COMMUNITY STORIES

Public Lab's contributions come from all over the globe.
Here are a few of the members that are providing tools for all to use.

ROCKETS.XIA, co-founder of Mushroom Cloud Makerspace
Location: Shanghai, China

Our project, Knowflow (github.com/KnowFlow/KnowFlow_AWM), is a water quality monitoring project funded by the Green Seed foundation and Mushroom Cloud Maker Space, with contributions from Shan He (another Public Lab member based in China). The Knowflow is still in development, but I think we will be able to use it to make water quality monitoring work easier.

MAX LIBOIRON
Location: Newfoundland, Canada

I direct the Civic Laboratory for Environmental Action Research, a feminist, anti-colonial laboratory that specializes in grassroots environmental monitoring of marine plastic pollution. I helped design BabyLegs, a low cost, open source, accessible marine trawl, for sampling waste (especially microplastics) in subarctic regions of Newfoundland and Labrador, Canada (see page 24).

MAGGPI AKA MAGGIE NORTON
Location: Washington, DC

My work involves adapting Public Lab's open hardware microscopes to image in infrared by swapping both the light and camera sensors. Applications include looking at insect circulatory systems and enhancing contrast of cell features.

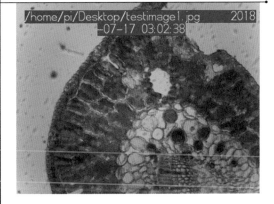

VJ PIXEL
Location: Brazil

I work with Rede InfoAmazonia, a group that has developed Mãe d'Água, an inexpensive monitoring system to analyze the quality of water for human consumption. We are establishing a monitoring network in partnership with communities in the Brazilian Amazon. ◉

ECONOMIA CIRCOLARE

Classical art style is given a 3D rendering in this 2018 exhibit from Maker Faire Rome.

Italians designs for a
carbon-free future
at Maker Faire Rome

Written by Jennifer Blakeslee

JENNIFER BLAKESLEE is the Maker Faire global program manager and has been a maker at Maker Faire and in the Bay Area for the last decade. She is also a mama of two little humans, and lives in Oakland, California.

Waste, regeneration, and sustainability are buzzwords of the Anthropocene, yet makers have been hard at work cracking these codes for centuries. As an abundance of materials grows into an abundance of trash, we're now seeing collaborations on innovative solutions that have "reduce, reuse, recycle" built into their DNA. Organized for the 7th cycle by the Rome Chamber of Commerce, Maker Faire Rome has spent several years elevating the work of these citizen scientists in a special pavilion dedicated to the circular economy. And, the event is going all out in an effort to be completely carbon free, prohibiting the use of plastic containers and relying on renewable energy sources across the production. Bursting with clever design and serious technology, the Circular Economy Pavilion, as well as six other curated topic areas, feature makers working on a diverse array of bioeconomy, greenbuilding, and greentech projects that get to the heart of our collective future.

Fast growing-low resource food production; open source parks and ethical social housing designs; hybrid roads and innovative cycling modifications; water and bioremediation techniques; and renewable fashions will all feature on the pavilion floors (Figure ). A new maker art initiative, which sees art as an engine for raising awareness and a vehicle for technological experimentation, illuminates the art-tech relationship and its sensitivity to environmental issues.

Some upcoming highlights: Joaquin Fargas uses robotics to investigate utopian proposals of

sustainability and human relationships; Mattia Casalegno, whose immersive works submerge participants in nature and gastronomy, will show his *Grass Roller* for the first time in Italy; and *Reflection on Life* by Anna Frants and *The Enlightened One* by Elena Gubanova, both from the Cyland MediaArtLab, represent unique ways of seeing life and loss in our relationship with nature will all be showing work. Many other artists such as Pier Alfeo, Simone Pappalardo and José Angelino, Matteo Nasini, Maria Grazia Pontorno, Martin Romeo, Lino Strangis, and Mat Toan also navigate new concepts of our relationship with the environment. Massimo Banzi, cofounder of Arduino and curator of Maker Faire Rome, has made the event a crucible for the *new industrial revolution*, calling himself "a happy witness" of a place where new enabling technologies and new models of innovation — and the people who create them — can come together.

Although Maker Faire Rome is a profoundly international event, here are a couple unexpected "local" projects looking beyond the "take-make-dispose" paradigm.

FRUIT TO FABRIC

Orange Fiber (orangefiber.it), founded by Adriana Santanocito and Enrica Arena (Figure **B**), has developed a method to extract cellulose from the byproduct of citrus processing, creating silk-like yarns and elegant, innovative, sustainable fabrics (Figure **C**) that have been picked up by fashion houses and fast fashion retailers. Santanocito, a Sicilian native familiar with the region's *pastazzo* (citrus waste) problem of more than 700,000 tons per year, discovered that it was possible to extract cellulose from orange rinds and spin it into yarn while tinkering in the labs at Politecnico di Milano in 2012. Feasibility studies, patents, awards, and crowdfunding followed and the company has expanded its operations and now operates out of a Sicilian juice factory, where they source their materials for free. As Arena noted last year, "Compared to existing man-made fibers from cellulose (from wood, hemp or bamboo) our fiber does not require dedicated yields alternative...but reuses a waste thus saving land, water, fertilizers, and environmental pollution." They'll be at Maker Faire Rome again this year.

MODERN CLASSIC

Although Vespa introduced its Elettrica model in late 2018, as Europe's largest market for two-wheeled scooters Italy will have plenty of gas guzzlers for years to come. **Motoveloci's Retrokit for Vespa** (motoveloci.it) replaces the engine of the traditional scooter with a battery powered driver without making structural changes to the moped's iconic design (Figure **D**). The electric motor fits perfectly into the engine compartment originally designed by Piaggio, the extractable battery is placed in the saddle (Figure **E**), and the petrol tap is replaced by an ignition block for the electric motor. The kit includes an aluminum alloy crankcase and motor with a maximum power of 6kW, a 1.5kWh removable lithium battery that can be recharged in three hours with a standard 230V socket, and an app for controlling driving parameters and mappings. The maximum speed of the Vespas can be brought up to 90km/h (about 56mph) by unlocking the limit via the app and Retrokit scooters have an urban autonomy of ~100 km (~62 miles) with energy recovery during braking.

Don't miss Maker Faire Rome — The European Edition, October 18–20, 2019. ◐

D

E

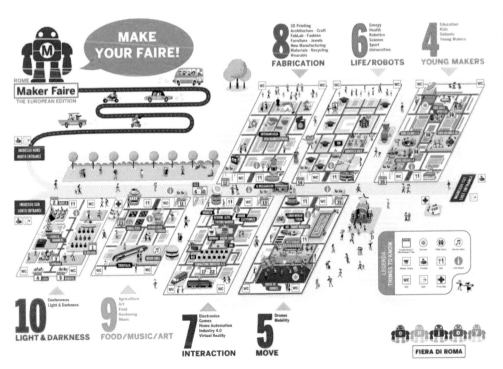

Courtesy of Orange Fiber, courtesy of Leonardo Ubaldi and Alex Leardini; Carola Ghilardi

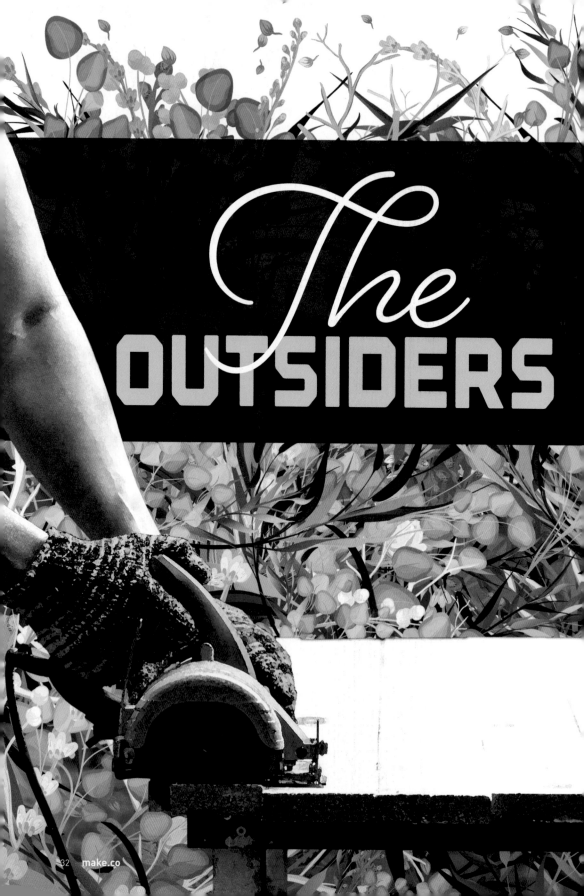

The
OUTSIDERS

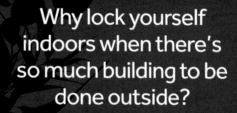

Why lock yourself indoors when there's so much building to be done outside?

There's something so simply gratifying about building outdoor projects. Perhaps it stems from our early need for shelter, having to mill lumber, bake bricks, and develop structural joinery to keep ourselves warm and dry. Or maybe it's from creating agricultural systems and hunting tools to satisfy our food cravings. But as solutions to those needs are well established now, our personal outdoor projects have primarily become recreational, and that's both fine and fun! To celebrate this, we've assembled a batch of enjoyable activities to both build and use outside. Now get off your chair, grab your tools, and go enjoy those blue skies. ●

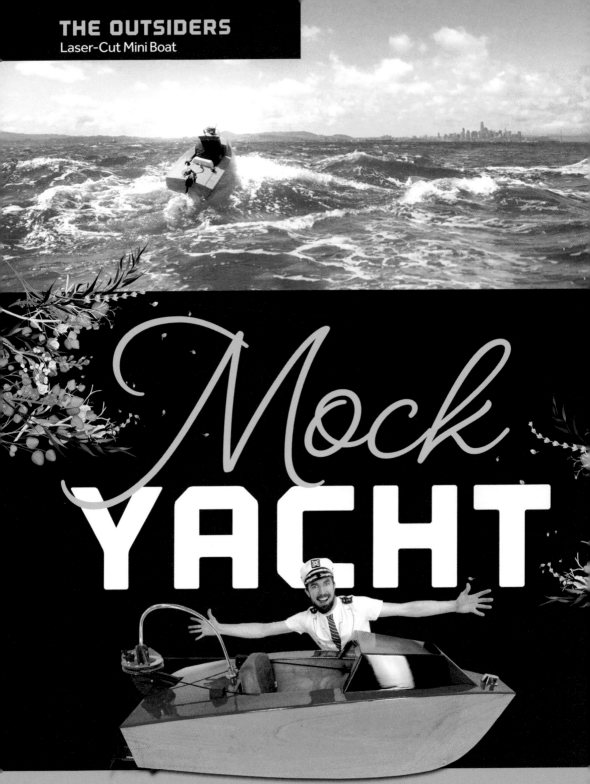

Mock YACHT

Explore the seven seas in this just-barely-one-person motorboat **Written by Josh Tulberg**

TIME REQUIRED:
100 Hours

DIFFICULTY:
Moderate

COST:
$900–$2,100

MATERIALS
See the full BOM at rapidwhale.com/mini-boat.php. Buy the plans and 3D files ($95), then you can DIY it all, or get a bare-hull kit ($950) with the pre-cut plywood, acrylic steering wheel, Delrin bearings, and gaskets, and the 3D printed parts.

» **Plywood, okoume BS1088 marine grade, 6mm thick, 4'×8' sheets (3)**
» **Acrylic sheet, ¼" × 21" × 15"** for steering wheel
» **Delrin sheet, ¼" × 10" × 7"** for steering bearings
» **Cork sheet, 1/16" × 10" × 7"** for gaskets
» **PVC pipe** for steering shaft
» **Rope, pulleys, quick-cleats** for "go-kart" steering
» **Electric trolling motor**
» **Trailer electrical connector, 4-pin**
» **Battery, 12V deep cycle, and charger**
» **Battery clamps and straps**
» **Wire, 8AWG and 12AWG**
» **On-off switch, heavy duty**
» **Bow light, red/green, 12V**
» **Fuse, in-line**
» **Voltage gauge** to estimate remaining battery life
» **Vinyl tubing**
» **3D printed parts: steering drum hub, pulley spacers, motor controller knob, etc**
» **Marine epoxy, 5:1 resin and hardener**
» **Wood glue and 5-minute epoxy**
» **Wood flour**
» **Cable ties**
» **Fiberglass tape, 3" wide**
» **Fiberglass cloth, 6oz weight, 5yds**
» **Liquid urethane foam**
» **Varnish**
» **Seat, carpet, and hook-and-loop tape**
» **Life vest** aka personal flotation device (PFD)
» **Emergency paddle**
» **XT90 connector** for a simple dead man's switch
» **Lanyard** to tie dead man's switch to your PFD
» **Screws, nuts, and washers, stainless steel**
» **Paint and/or stain (optional)**

TOOLS
» **Laser cutter (optional)** minimum 900mm × 600mm cutting zone
» **3D printer (optional)** Print the 3D files or send them out to a service.
» **Files, sandpaper, clamps, hacksaw, drill, screwdrivers**
» **Soldering iron**
» **Stir sticks, cups, brushes, and gloves**
» **Icing bags**
» **Fillet tools**

JuliAnn Tulberg

Have you ever wanted to build a boat? How about a really (really) small one? If so, you may be interested in building this 6-foot mini boat. Some highlights:

• Super darn cute and a joy to ride
• 100% electric propulsion from an outboard trolling motor
• Made from precision laser-cut components
• Interlocking assembly means no jigs required
• Simple, quick cable-tie and epoxy construction
• Interior bulkhead design — floats when flooded
• Steers with a beefy Plexiglas steering wheel
• Low and comfortable seating position
• Surprisingly stable flat-bottom hull design
• Reliable and accurate steering geometry
• Convenient cubby storage (above dash)
• Additional behind-seat storage
• Designed to fit a 6'2" tall rider 200lbs or less

I was inspired to build my own mini-boat after seeing a few others, most notably Paul Elkins's 8-foot *Little Miss Sally*. I designed mine to be a 6-footer, the smallest I could make it with a proper center of gravity that let my legs be straight.

I wasn't too keen on cutting plywood by hand, so I designed it for modern manufacturing techniques. I started with 6mm marine-grade plywood and then cut each panel with a laser cutter into an easy-to-assemble shape. I also 3D printed a handful of small parts for ease of assembly. Shameless plug: I sell bare-hull kits or digital plans for making your own at rapidwhale. com/mini-boat.php. As far as I know, this is the first-ever boat cut with a laser. A lot of boats are cut with a CNC router, which typically has a much larger (4'×8'+) cutting zone than a laser. But a mini boat, with small and intricate panels, is perfect for a modest ~3'×2' (cutting zone) laser.

The boat is assembled using cable ties, which hold the pieces together in preparation for epoxy. This is called a "stitch-and-glue" construction technique, although typically the stitching is done with copper wire. I'd heard of people using cable ties, but sticking a rectangular profile (cable tie) through a circular hole (drill bit) didn't sit well with me. Fortunately, a laser allowed me to cut the sharp corners required to perfectly match the profile of a cable tie.

A

B

CAD

C

DESIGNING A BOAT

As with most of my projects, I spent more time designing than actually building. My concept sketch started on a sticky-note (Figure Ⓐ). That note sat in front of my desk for a few months as motivation to actually start such a daunting project.

Eventually I imported a variation of that sketch into SolidWorks (Figure Ⓑ) and designed around it. I simulated the material and water displacement in order to determine the vessel's center of gravity and center of buoyancy. I'm no boat expert, but I had read that the relationship between center of gravity, center of buoyancy, and hull shape is critical for determining (or estimating, in my case) boat stability.

From there I laser cut a 1:2.2 scale model to test in my friend's hot tub, which makes for an awesome choppy-water simulation, by the way (Figure Ⓒ). I loaded up the scale model with scale weight. The model floated, but it wasn't as stable as I was hoping for, so I revisited my CAD model. I needed to move the center of gravity forward, and to slightly widen the hull shape. Sounds simple, but that meant 40+ more hours of design edits and file prep. Ouch.

BUILD THE SUB-ASSEMBLIES

The inner sub-assemblies are simple — the seat base, steering wheel, and steering drum. You'll glue these together, let them dry under weights, and file away any excess glue (Figures **D** and **E**). Binder clips make good clamps here.

Cut the motor shaft to length such that the total exposed shaft length is exactly 22.35". Use the 3D printed hub as a guide for drilling the holes (Figure **F**). Remove the protective metal sleeve that was protecting the wires, and ensure the only loose item on the motor shaft is the motor-mount itself.

Epoxy the steering drum hub to the wooden spokes. Quickly epoxy the hub/spoke assembly into the drum assembly (Figure **G**). You may want to use a little hammer force here.

Drill an access hole (~½" diameter) into the assembly so that you can fit the bottom bolt. Press-fit the thrust bearing into the 3D printed steering drum hub. Attach the steering drum to the motor shaft without pinching the internal wires.

PREP FOR ASSEMBLY
EPOXY THE PUZZLE JOINTS

Some of the larger boat panels are split into multiple pieces. You'll epoxy these together to create the larger panels. Fear not, these puzzle joints end up being just as strong, if not stronger, than the rest of the plywood (Figure **H**).

Run a small amount of 5-minute epoxy (or marine epoxy) along both edges of the puzzle joint. Use a flat board as a press (wax paper can be used to keep the epoxy off the flat board) and stomp on the board to ensure a flush press-fit of the puzzle joint. The press-fit should be tight enough that you don't need to use any serious clamps. Repeat this for all puzzle joints.

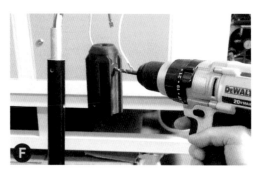

> **CRITICAL:** When gluing the nose pieces (of the hull-sides), triple check that they will be glued such that the bottom of the boat is *flat* (not curved). This is a *flat-bottom* boat. It is easy to make this mistake.

GLUE MOTOR MOUNT TO TRANSOM

This is where the removable trolling motor will tighten down onto your boat, so it will need to be

THE OUTSIDERS
Laser-Cut Mini Boat

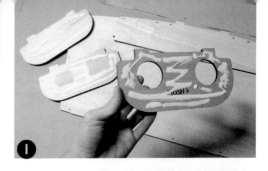

reinforced. Apply liberal amounts of wood glue and stack the pieces accordingly, making sure the cable-tie slots are all aligned. The piece with the two holes (Figure **I**) is the topmost piece to be stacked. Put it on last. Be sure they're all perfectly aligned and let them dry under weights. Your battery makes a very good weight.

SAND PUZZLE JOINTS FLUSH
Sand the joints after the epoxy has cured. Don't go crazy with the sanding, you're just trying to remove the excess epoxy from the visible surface.

CUT FIBERGLASS
Use the newly formed pieces to trace and cut the fiberglass accordingly (Figure **J**). Cut the fiberglass right on the line because you will not want this fiberglass to wrap around any edges (that's what the 3" fiberglass tape is for).

ASSEMBLE THE BOAT
BEGIN CABLE TIE ASSEMBLY
Knock out the large access circles in the bulkheads. Using the cable ties, attach the bulkhead plywood to the base of the hull first. Next attach the sides of the hull, and then the transom. Leave the cable ties loose so you can still make adjustments (Figure **K**).

> **NOTE:** Make sure the cable ties are oriented such that the heads are visible *after* assembly.

Carefully form the hull shape by bending the plywood and attaching cable ties accordingly (Figure **L**). If you bend too quickly, or unevenly, then you may crack the plywood near the kerf cuts. You can repair any cracks with epoxy later.

Next, attach the dashboard. You'll also want to glue the two small pieces to the dashboard, and possibly cut the profile of the dead-man's XT90 connector before assembly (Figure **M**).

Attach the piece that's similar in shape to the dashboard to the bottom of the deck. Then attach the deck (Figure **N**). Finally, attach the small pulley block pieces connecting the dashboard to the similar shaped piece.

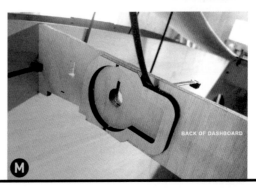

BACK OF DASHBOARD

N

Now, begin cinching down all the cable ties. The wood pieces interlock, and you're left with a tight edge (Figure **O**) if you're doing it right.

Snip the dangly bits off most of the cable ties, so that they don't get in the way as you begin the next step (Figure **P**). At this point your entire boat should be assembled, minus the plywood windshield piece that will go on later.

EPOXY FILLET ALMOST EVERYTHING

Mix epoxy and wood flour until you get a peanut-butter consistency, and use the icing bags to apply. The proper consistency takes a lot of wood flour.

Channel your inner contortionist because it's time to fillet all of the corners *inside* the bulkhead (Figure **Q**). This is by far the most challenging part of the construction. You will have to apply the fillets blind because unfortunately there isn't enough room to get a head *and* arm into the bulkheads.

Apply fillets to the inside corners of the boat (Figure **R**, following page). Run a wooden fillet tool over as many as you can to make them perfect. At this point nearly every cable tie and corner on the inside of the boat should be covered in an epoxy fillet.

FILL BULKHEADS WITH FOAM

The bulkheads should be airtight once construction is completed. The foam is just there

O

P

Q

Josh Tulberg, JuliAnn Tulberg

R

S

T

U

V

to ensure that even with a leak, the bulkheads will still displace water.

Epoxy the small wooden tabs into the back of the bulkhead hole, and the large wooden circle into the bulkhead. Align the grain if you want to be fancy. Be sure to epoxy all around the circle so the foam won't push it out when it expands.

Quickly mix the foam (Figure **S**), only as much as you need. If you purchased the same foam as in the BOM, then exactly half of that is required for each bulkhead.

Pour the foam. It will quickly start to expand, so rock the boat such that the still-liquid foam gets distributed to fill the entire cavity evenly. The foam gives off a good amount of heat and creates a good amount of pressure. If you do everything right, all of the bulkhead plywood should be warm, and the excess foam should exit out the small hole in the wooden circle. Wait for it to dry.

Seal the bulkhead by carving out excess foam and gluing the small plywood circle into the hole.

CONTINUE ASSEMBLY

Snip the cable-tie heads off the floor of the cabin (the place where you sit), so that you can more easily run a fillet bead. The now-cured epoxy fillet that you laid along the inside of the bulkhead will hold this piece in place.

Apply a fillet along the base of the floor, where you just cut the cable-tie heads (Figure **T**). As usual, clean it up with a fillet tool. At this point every single corner inside the boat should be covered in a nice-looking epoxy fillet.

Next, cable-tie the plywood windshield together. Keep these fairly loose and ensure that the cable ties' heads are visible (outside) after assembly (Figure **U**).

Attach the windshield to the deck with cable ties. It's easiest if these cable ties are already fed through the slots in the deck before placing the windshield (Figure **V**).

Cinch down all the cable ties. It helps to use needle-nose pliers. Epoxy fillet all inside corners of the windshield area, even those inside corners where the windshield meets the deck. This ensures a permanent bond to the rest of the boat. Let the epoxy cure.

Josh Tulberg

ADMIRE YOUR WORK THUS FAR

You just turned a pile of wood into a shape that resembles a boat!

WATERPROOF THE BOAT

You'll do this the traditional way using fiberglass and epoxy, followed by varnish (Figure **W**) and, if you wish, paint. Go to the online plans at makezine. com/go/mini-boat to follow my specific notes.

FINAL ASSEMBLY AND WIRING

Now that the body is completed, we'll get the mechanics and electrical together to make it move.

Begin assembly by bolting on the rear pulleys (Figure **X**). The 3D printed spacers keep the pulleys level when mounted on the curved deck, and the smaller spacer goes under the deck (not visible) to keep the nuts level. The cover is held on with the one innermost bolt, so tighten that bolt up

last, after you place the cover. There are mirrored pieces for port and starboard sides, so no two pieces are identical. Pay attention to how you are assembling them and don't mix up the covers.

Next, bolt on the front pulleys (Figure **Y**). Bolt them on staggered: if you bolt one closer to you, bolt the other one on the other side further from you. You need the pulleys to be staggered so that the rope runs through them and around the steering shaft smoothly. Use fender washers on all spots where washers touch wood.

Bolt the quick-cleats onto the steering drum, then bolt on the Delrin steering shaft bearings. Float the steering shaft in them as you tighten down. That way you can rest assured the two shaft bearings are concentric to the shaft after tightening.

Drill holes in your steering shaft according to Figure **Z** and bolt the 3D printed steering hub

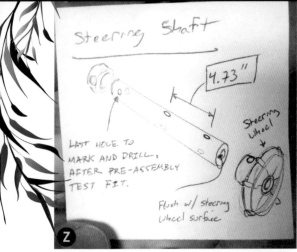

Steering Shaft

4.73"

LAST HOLE TO
MARK AND DRILL,
AFTER PRE-ASSEMBLY
TEST FIT.

Steering
Wheel

Flush w/ steering
wheel surface

Z

Aa

Bb

Cc

onto the shaft and bolt the steering wheel onto the hub. Install the shaft into the boat, then bolt on the final 3D printed steering shaft piece to hold it all in (Figure Aa).

Press-fit the motor controller. It goes in from behind the dash, and you may need to spend some time filing excess epoxy to get it to fit perfectly.

Next, press-fit the 3D printed motor controller knob onto the motor controller shaft (Figure Bb). This holds the motor controller in place.

Attach the straps, seat, and hook-n-loop strips to the seat base (Figure Cc). Be sure to thoroughly clean the seat base where you plan on attaching the peel-n-stick adhesive.

WIRING

First things first: Review the wiring diagram (Figure Dd).There's a higher-res version in the online plans.

Start by wiring the motor. It's never a good idea to use smaller wire than stock, and this motor comes with 10AWG wire, so I suggest transitioning to 8AWG wire since we need to lengthen it to reach the dashboard.

Run the motor wires through vinyl tubing; ½" ID, ⅝" OD was the ideal size (Figure Ee).

Connect the motor wires to the trailer hitch connector. I didn't need to solder directly to this connector, I was able to tin the wires then file 'em down to fit inside the holes (Figure Ff). If you strip a bolt, or can't get the wires to fit, then I wouldn't hesitate to just solder directly.

Next, create your dead man's switch. Use a proper XT90 connector, which is rated for up to 90 amps; this is fine for a trolling motor that calls for peak current of 40 amps. My original Deans connector (Figure Gg) failed; this is a better option.

Begin creating the cabin wiring harness. You'll

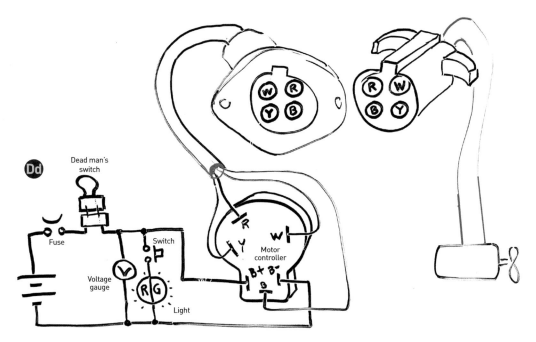

Dd

Dead man's switch

Fuse

Voltage gauge

Switch

Light

Motor controller

R
Y
W
B+ B−
B

Ee

want to measure, perhaps with a string, and cut your wires accordingly.

Install the bow light with its gasket (Figure **Hh**, on the following page). Then install the wiring harness you created. I ran the wires through a large-diameter tube and hot-glued the ends in an attempt to waterproof them. It's not necessary though, you could just run cable ties every 6" or so.

Wire in the fuse. It should be as close to the battery terminal connection as possible (Figure **Ii**, following page). I mounted mine up under the deck where the trailer connector is located.

Connect the wires to the in-dash motor controller (reference the wiring diagram), and cable-tie the wiring harness out of the way (Figure **Jj**, following page). Alternatively, you could drill a hole in the farther wooden piece

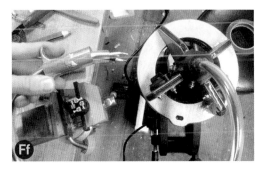

Ff

Gg

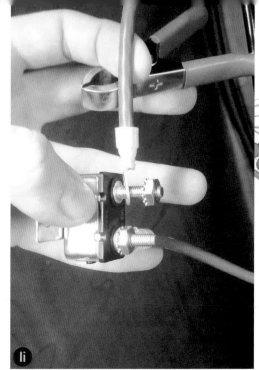

Josh Tulberg, JuliAnn Tulberg

to run the wires through that. Coat the edges of any hole you drill with 5-minute epoxy in an attempt to maintain waterproofing of the wood. The wiring harness should look clean and tidy if you did a good job with cable management.

FINISHING ASSEMBLY

Install the seat, using the hook-and-loop tape. Make sure to clean the floor of the boat to ensure the peel-n-stick adhesive on the tape will stick. I installed my seat such that in the fully open position, it was just barely touching the wood behind it. I had read that these seats don't like a lot of weight reclining on them, so I think it's a good idea to have it butt up against the wood

support. Mark the perfect location so that you can place it there again with adhesive. You won't get a second shot at this, and the industrial-grade hook-and-loop is pretty much a permanent mounting solution.

Finally, run the steering rope. It's critical that you run the rope as pictured in Figure Kk. If you deviate from this, then you'll either limit your steering ability, or you'll have backwards steering.

That's it! You should now be ready for the water. Document and enjoy.

Get more information, pictures, and videos of the Mini Boat on my site:
rapidwhale.com/mini-boat.php.

JOSH TULBERG is an industrial designer and recreational engineer running his own one-man prototyping shop.

Written by Toni Klopfenstein

Smart BIRD FEEDER

This high-tech roost doesn't just nourish your winged friends — it identifies and photographs different bird species (and will deter squirrels, too!)

TIME REQUIRED:
Two weekends

DIFFICULTY:
Moderate

COST:
$150–$250

MATERIALS
ELECTRONICS
» **Coral Dev Board** Mouser #212-193575000077
» **Coral Camera with 24-pin FFC cable** Mouser #212-193575000442
» **Speaker, 4Ω 3W** Adafruit #1314
» **MicroSD Card, 32GB**
» **USB-C power supply, 5V 3A**

BIRDHOUSE + ENCLOSURE
» **Acrylic sheet, clear, 6mm thick: 400×400mm (4) and 450×350mm (3)** Use clear acrylic so the camera can see the birds.
» **Nuts, M3 (58)** McMaster-Carr #91828A211
» **Nut, ¼-20** McMaster-Carr #91845A029
» **Screws, M2×6 self-threading (14)** McMaster-Carr #96817A844
» **Screws, M3×16 truss head (62)** McMaster-Carr #92467A433
» **Hinges (2)** McMaster-Carr #1603A27
» **Screws, #8 to #10, 1½" length (3) (optional)** such as Lowe's #9×1½" lattice screws
» **3D printed parts** Download the free 3D models at github.com/google-coral/project-birdfeeder.
» **Zip ties, tape**

TOOLS
» **Screwdrivers: 1.5mm flat, #1 Phillips**
» **Drill with 9/64" straight bit**
» **Nut driver, 5.5mm**
» **Torx driver, T6**
» **Diagonal cutters**
» **3D printer**
» **Laser cutter**

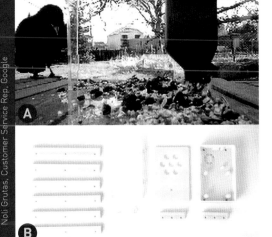

Noli Grutas, Customer Service Rep, Google

The Smart Bird Feeder leverages Google's Edge TPU integrated circuit on the Coral Dev Board to identify which animals are using the bird feeder (Figure **Ⓐ**). The Edge TPU can process the camera feed through an on-device machine learning MobileNet quantized model without needing remote computing power. The model classifies whether there's a visitor or not, and if so, what species it is. If an animal is detected, snapshots of the camera feed and the model results are stored on a microSD card. Additionally, the feeder can be configured to trigger a deterrent sound if a squirrel is identified. You can even use the photos gathered from the bird feeder to run on-device model training, and improve the model over time to more accurately identify local birds.

This project includes directions for building a custom birdhouse to hold the feeder and is designed around the Coral Dev Board. However, feel free to use your own custom birdhouse and enclosure design. You can also use a Raspberry Pi and a Coral USB Accelerator in place of the Dev Board for a more economical build.

TIP: If you're building a custom birdhouse, remember to design your enclosure to allow enough light for the camera to function properly.

NOTE: The build instructions assume you have already completed the initial configuration of your Dev Board. If you haven't, please refer to the Getting Started Instructions at coral.withgoogle.com.

GATHER MATERIALS
Download all the necessary files for this build (code, Edge TPU models, and 3D models) at github.com/google-coral/project-birdfeeder. 3D print the electronics enclosure, feed bottle, brackets, and optional mount plate (Figure **Ⓑ**). For our version, we used an SLA printer.

CONNECT THE ELECTRONICS
Use the flat-blade screwdriver to connect the speaker to either the left or right speaker port on the Coral board. Then, attach the camera to the board: flip the latch up on the board's CSI camera connector, slide in the flex cable, and flip the latch down. Repeat this process with the other end of the cable on the Coral camera.

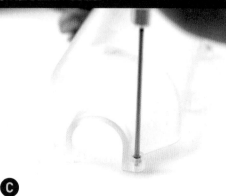

C

D

E

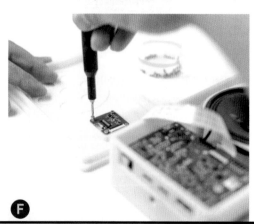

F

MOUNT THE ELECTRONICS

Insert the M3 nuts into the enclosure bottom, brackets, and feed bottle (Figures **C** and **D**).

Carefully line up the board and speaker inside the electronics enclosure (Figure **E**) and screw them into place using M2 screws.

Use 4 M2 screws to attach the camera onto the lid, taking care to avoid damaging the camera cable (Figures **F** and **G**).

Once all the parts are secure in the enclosure, connect the lid to the base. Make sure the pieces are flush once screwed together (Figure **H**).

CONFIGURE THE ELECTRONICS

Plug in the SD card and copy the directory **project-birdfeeder/code** you downloaded to the board's home directory.

Run the **config.sh** script on the Dev Board to configure storage and ensure you have the proper Python libraries installed.

Test the **birdfeeder.sh** script by holding up different household items to the camera. You will see the inference rate, inferred results, and the associated score, as well as the filename of any images captured, in a terminal, as such:

```
Inference: 4.38 ms, FPS: 12.60 fps
Sunglasses, dark glasses, shades, score =0.53
Sunglass, score = 0.17
Frame saved as: sdcard/img-0087712102.png
```

> **NOTE:** This script can be configured to run different models. Edit this file to change which model your birdfeeder uses, and to run the birdfeeder in training mode or deterrent mode.

BUILD THE BIRDHOUSE

Use the downloaded CAD files to cut the acrylic pieces for the birdhouse (Figure **I**).

Assemble the main house. Use a 7-nut corner bracket and 6 M3 screws to connect one edge of the front panel to one edge of the side panel. (Figure **J**). Repeat for each corner of the birdhouse (Figure **K**). Use the two short brackets and 8 M3 screws to connect the bottom piece to the rest of the house (Figure **L**) for your final result (Figure **M**).

Next, attach the electronics box. Run the power cord through the hole in the house bottom, and plug it in to the Dev Board (Figure **N**). Secure

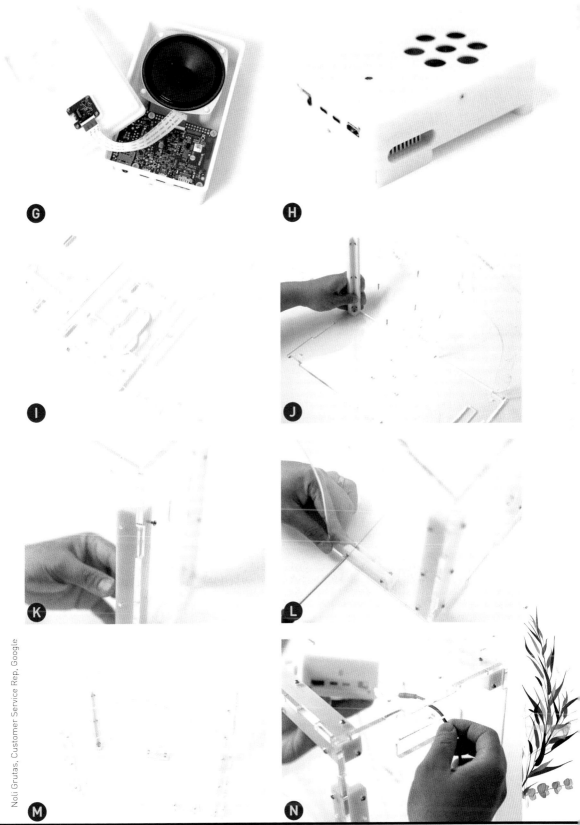

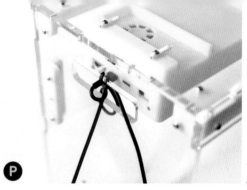

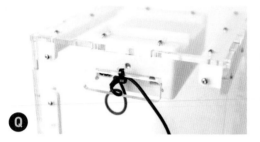

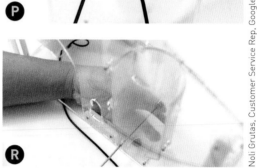

the electronics box to the 4 slots on the right wall using 4 M3 screws (Figure **O**, page 50). You can adjust the box up and down in the slots to adjust the camera view.

Secure the power cord. Make a loop with the cord and secure this to the secondary hole in the base with a zip tie (Figure **P**). This provides strain relief and prevents power getting disconnected. Trim off the excess zip tie (Figure **Q**).

Secure the birdseed bottle inside the back panel using 6 M3 screws (Figure **R**).

Connect the roof: Connect the slotted roof piece to the house using 2 long brackets and 8 M3 screws (Figure **S**). Using the hinges, secure the remaining roof piece to the slotted roof piece (Figures **T** and **U**).

> **NOTE:** If you're concerned about rain getting into the birdhouse, add a piece of tape over the roof seam.

INSTALL THE BIRD FEEDER OUTSIDE

Screw the base of the feeder or the optional mount plate into a tree or post using the #10 screws (Figure **V**).

If using the optional mount plate, align the nubs on the mount plate with the holes on the back of the bird feeder. Slide the feeder down until secure.

Now add birdseed. Check with your local birdwatching society or wildlife agency about the appropriate feed for your local area. Different species require different foods, so you may need a variety of food to attract a mix of birds.

> **NOTE:** It will likely take a few days for your local birds to discover the feeder and become comfortable enough to approach it. To encourage them, ensure that the feed stays dry, and that predators cannot access the feeder easily. Moving the feeder to a different location may also help.

COLLECT DATA

Plug in your power supply and verify you see the power LED on your board. It's working now. Let the feeder run for a week to gather enough data.

The SD card will store any captured snapshots along with a **results.log** file. This file should show what species the bird feeder model identified, the confidence of the model results, and what the associated image ID is, such as:

```
2019-04-11 12:56:00,713 -Image:
0012338723 Results: [('Corvus
cornix (Hooded Crow)', 1.6796875),
('background', 0.9453125)]
```

Verify that the model results match with the associated camera image. If your model isn't accurately identifying the animal visitors, it's time to train your own custom model on device!

RETRAIN A MODEL

Create a dataset from your captured photos. Group your collected photos into directories named after their respective identifying labels (for example, *background*, *pigeon*, and *rabbit*).

Run the script **classification_transfer_learning.py** on an embedding extractor to create a custom retrained model based on your new dataset. Next, update the **birdfeeder.sh** script to use your new **.tflite** model and **labels.txt** file, and remove the training flag.

Now test your new model out. You'll see something like:

```
2019-04-15 19:07:15,075 - Image:
0087713445 Results: [('fox squirrel,
eastern fox squirrel, Sciurus
niger', 6.7578125)]
2019-04-11 19:07:21,964 - Deterrent
sounded
```

If you aren't getting the accuracy you want from your model, try using a different data set or using a different embedding extractor to retrain the model.

> **NOTE:** For more detailed information on the retraining process, please visit coral.withgoogle.com/docs/edgetpu/retrain-classification-ondevice.

GOING FURTHER

If you want to enhance or modify your feeder, consider adding a motion detector or other environmental sensor to only trigger the camera at specific times.

To capture different species of birds, you can deploy the feeder to a more remote location by adding a solar panel, charger, and appropriately sized battery to power the electronics. ◉

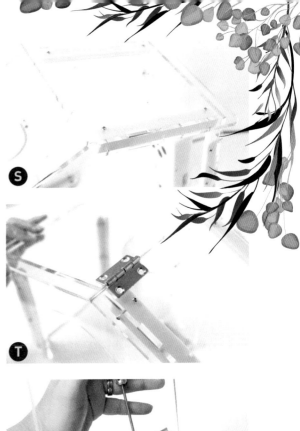

S

T

U

> **NOTE:** Birds prefer to be farther away from the ground, and often prefer tree coverage or other protection while eating. Hard-to-reach locations also help deter squirrels.

V

TONI KLOPFENSTEIN is a Developer Advocate at Google who focuses on IoT.

Longboard
CNC SUP

Use a CNC router to build a beautiful, hollow wooden stand-up paddleboard

Written and photographed by Sonny Lacey

TIME REQUIRED:
Several Weekends

DIFFICULTY:
Intermediate

COST:
$150–$400

MATERIALS
» **Plywood, ¼"×48"×96" sheet** for the skeleton and fin. Marine grade is great, but you can also just use the most knot-free plywood you can find.
» **Wood strips, about 3"×¼" thick, 12' lengths (20)** for the covering, about 64 sq.ft total. Cedar is a popular choice. You can also cut plywood strips, or recycle flooring or fence boards, but these take more work.
» **Extruded polystyrene (XPS) foam, 2"×48"×96" sheet** such as "pink foam" or blue insulation board, for the "NeverSink" rails.
» **Wood blocks, 24"×3½"×2½" thick (2)** for nose and tail blocks; solid wood or glue-ups
» **Surfboard vent plug with leash attachment** greenlightsurfsupply.com
» **Waterproof wood glue** e.g. Titebond III Premium
» **All-purpose construction adhesive** such as Titebond PL Premium
» **Epoxy, 2:1 ratio marine "no-blush" type (1gal)** Basic Marine No-blush from Progressive Epoxies
» **Fiberglass cloth, 4oz plain weave, 40" × 9 yds**
» **"Spar" varnish, water based, 1qt** Cheap Last-n-Last or fancy marine varieties will work.
» **Paint suitable for plastics** e.g. Krylon, for rails or other areas as you choose. Avoid the high-end "topside" marine paints.
» **Plastic zip ties and/or wire**
» **Nitrile or latex gloves** Buy a pack of 100.
» **Packing tape and painter's tape**

TOOLS
» **Dust mask** or similar. I recommend a cartridge-type respirator that covers that mouth and nose.
» **Eye and ear protection**
» **CNC router, at least 29" wide** I use my Maslow upright CNC — see Step 1. Access a CNC at a makerspace or send the files to a cutting service.
» **Computer with vector graphics editor and G-code generator** such as Inkscape and MakerCAM, both free
» **Circular saw with 8" blade minimum**
» **Jigsaw**
» **Pull saw or compass saw**
» **Straps and/or plastic wrapping**
» **Clamps** varying from 2" to long bar clamps
» **Rope, about 20'**
» **Hand plane** such as a simple block plane
» **Sanding blocks and sandpaper** 80–440 grit
» **Sawhorses (at least 2)**

OPTIONAL BUT VERY HELPFUL:
» **Belt and orbital sander, power planer, Surform "grater" type planer, table saw, bandsaw, miter/chop saw, hot wire foam cutter**
» **A helpful partner** who can occasionally help hold, flip, or fetch things like clamps or coffee

---

Wha's SUP, water lovers?

A stand-up paddleboard (or SUP) is one of the newest incarnations of the venerable surfboard, and we're going to build one that has all the wonderful qualities of the old wooden longboards. In fact, this 10'5" longboard SUP can do double duty as both a paddle craft and a wave-rider.

I grew up around boats and the beach. The surfboard is a natural for attracting any young beach kid's interest. After hurling around some awfully heavy solid wood boards in the past, modern materials and techniques drew me back in. As I became interested in building lightweight sailing boats, I started to fuse modern composite technology with old-school wood. This CNC SUP project pulls in the best influences from drastically different worlds.

At the core of this CNC SUP longboard is a "rib and spine" skeleton that makes it hollow yet strong. Using a CNC router and the vector images that I provide, you'll be able to create this internal frame in a few quick steps. The covering of the board and the final finishing are phases of the project where you can go in any number of directions. I'll show you one way, and will hint at others. The beauty of making a surfboard is that there are infinite varieties in the final covering.

Finally, this SUP has excellent safety. It uses what I call a "NeverSink" system for its rails (the curved edges of the board) which result in a board that can survive a fracture and yet never fully sink!

BUILD YOUR CNC'D WOOD STAND-UP PADDLEBOARD
1. CNC THE SPINE AND RIBS

There are two SVG files: *ColdwaterLongboard2_nolabels.svg* and *ColdwaterLongboard2_labels.svg* (download at makezine.com/go/cnc-sup). You'll use the *nolabels* file to create your G-code. This file has the document set at 4' high by 8' long, for a full sheet of plywood oriented in a landscape manner on my Maslow CNC router (see *Make:* Volume 68, page 42). You can rearrange the parts to fit smaller routers.

The *labels* file contains all the same parts, but they're labeled (Figure **A** on the following page); print this out as a handy guide for inserting the ribs into the spine. Alternatively, and if you have

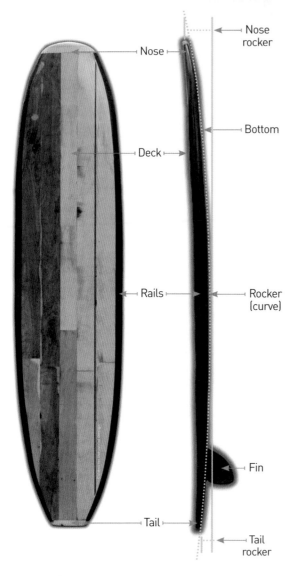

Nose rocker

Nose

Bottom

Deck

Rails

Rocker (curve)

Fin

Tail

Tail rocker

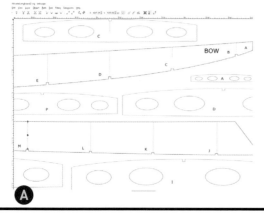

(A)

Z-axis (depth) control, you may decide to actually incorporate the labels into your cut paths; set their plunge depth to only 1/16" or so.

The SVG files have the ribs outlined in green and the two halves of the spine outlined in red. Short vertical paths indicate the slots, which you may have to adjust when creating your G-code if you've chosen a thicker or thinner piece of plywood for your parts. So, if your plywood sheet is only 1/4" thick, and your router bit is 1/4" thick, then on the slot paths, you'll want to ensure that the router is performing a *follow* operation and not an *outline* operation. If your plywood is thinner than your bit, then you'll have to cut the slots by hand; no biggie!

You'll notice that there are ellipses inside the ribs and spine; these are weight-saving holes in the structure, so these cutouts can be discarded.

NOTE: The SVG profiles do not indicate tabs. Tabs are nice to have, especially if you have a vertical CNC such as the Maslow; they help hold the pieces so they don't fall down on your final pass and meet with the spinning router bit. Usually, you can set up tabs in your G-code environment, such as MakerCAM.

Once you're good with the SVG files, create your G-code and load it into your CNC platform; it's time to cut! Load one standard 4'×8' sheet of plywood into your machine and safely cut the parts out (Figure **B**). After cutting, and especially if you used tabs, clean up the pieces with a rasp or sandpaper and you're on your way to assembly.

2. ASSEMBLE THE SPINE

The spine or *stringer* of the board is made from the two parts outlined in red in the SVG files. You'll "butt scarf" these two pieces together – glue the angled *scarf joint* together, and reinforce it with two small *butt blocks* of scrap wood as you would an ordinary square butt joint, for extra strength. To do this, lay the spine on a flat surface and place the diagonal cuts against each other to check for fit. Next, place an overlapping piece of wood behind it, glue all three pieces, and then glue an overlapping scrap of wood on top, using a weight as a clamp. Use a good, waterproof wood glue such as Titebond III and let the assembly dry for at least 8 hours.

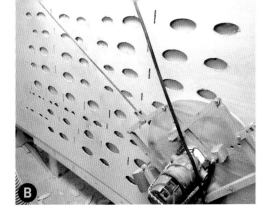

3. GLUE THE RIBS IN

Once the spine is assembled and the scarf joint is dry and strong, start gluing the ribs into their respective slots, using waterproof wood glue in each joint (Figure **C**). You'll want to do this on a table, in two phases — first the forward (*bow* or *nose*) half of the spine and then the rear (*stern* or *tail*) half — because the bend (*rocker*) of the spine makes it difficult to glue all the ribs perpendicular at the same time. To glue them all in one pass, you could make short supports and temporarily hot-glue them diagonally between each rib and spine.

Figure **D** shows all the ribs in place; on top is my "sight batten" — just a long, thin strip of material I use to align the whole board or to help make fair curves.

4. CREATE RIB RAIL STRINGERS

Rip 6 long strips of plywood 8' long by ½" wide. You can find this wood at the top and bottom of the plywood sheet you used to cut out the ribs and spine. These don't have to look great; they're placed on the outer edge of the ribs, and help support the foam rails that will come later. You can zip-tie them or wire them to the ribs while the glue is drying (Figure **E** on the previous page). Align these *rail stringers* toward the top third and bottom third of the rib edge, and overlap them so that each rib has at least two rail stringers on its edge (Figure **F**).

5. LOFT THE FRAME AND MEDITATE (OPTIONAL)

If you're able, run some twine through one of the holes in the spine near the bow and another piece of twine through a hole near the stern, and hang these from a rafter or ceiling. Now hoist the skeleton up to about eye level (Figure **G**). This is a good way to see if the frame is balanced.

If you need to tip the board one way or another, tape a small piece of wood near the outer end of a rib that must come down (Figure **H**); you can move the weight inboard (toward the spine) to fine-tune it. Once you're happy, glue it to the inside of the rib where it won't interfere with planking or rails.

Lofting the frame is a momentous occasion: You can see that this is really a creature made for a liquid medium. You can imagine how it will maneuver and feel. Sit back and meditate on how your board will look.

6. LAY OUT YOUR PLANKING

In most all cases, your planking will be 3"-wide clear pieces of lightweight wood, such as cedar. Decide now whether you'll do it in straight strips that run from bow to stern (longitudinally), or diagonally. You could also try bending large sheets of wood to the curvature of the board in both directions, but this can be quite tricky.

You'll plank the bottom first, and then the top. Planking can be done either off the frame or on the frame. Figure **I** shows a board I made with a plywood bottom skin, and a diagonal cedar top skin, planked on the frame.

Planking off the frame is easier. You'll want a large, flat surface where you can glue all your planks together for the bottom, and then again for the top.

Regardless of how you choose to do it, make sure you can clamp pieces together tightly enough to get a good bond; this may mean doing only a few boards at a time. Long bar clamps are great for this, as are "come-along" ratcheting straps.

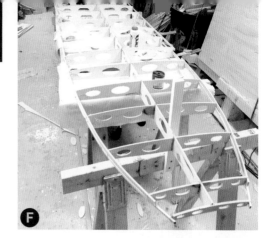

F

G

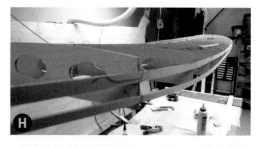

H

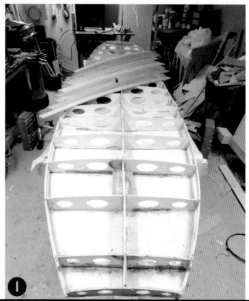

I

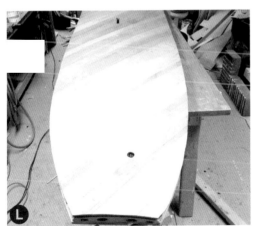

Packing tape can even be used if you get it tight enough.

Planking can even be done with strips of plywood, but be aware that when plywood is sanded to match the curvature of the frame, it will create "striping" where the high points are sanded down. Still, the interesting look of sanded plywood high spots may do it for you; I think it looks fine!

7. FIGURE OUT YOUR FIN

Once your bottom boards are all glued together edge-to-edge, you'll need to make a decision: fin box or no? A fin box is a plastic box with a slot that accepts a standard removable center fin. The box is placed in the underside of the board via a slot cut from the outside. Since the spine of the board runs down the center, you'll have to cut into the spine, or get creative and mount the box off-center.

If you don't want to go through with a fin box, you can make a nice fin out of scrap plywood and simply epoxy it on later.

> **TIP:** Plan to make your fin while you're covering the board in fiberglass and have a day to spare.

8. PLANK THE BOTTOM

If you've created the bottom skin off the frame, now's the time to glue it on. Take a look at the board. Whether you're planking piece-by-piece on the frame or in one go, you'll want to make sure that the board is not developing a twist. To knock out twist, prop up the ribs with support blocks on your table while you clamp the ribs to the planks, and place large weights on top as you're gluing the planking on.

9. MAKE THE VENT BLOCK

A hollow surfboard needs a vent to let air escape and equalize with the outside air, during temperature or altitude changes. The normal vents are small plugs about 1" wide and 1½" deep. Cut a small block of scrap wood about 3"×3"×1½" deep and drill out a hole the size of your vent plug. You'll put this block on the inside of your top skin. First lay the skin on the board and measure where the vent box will go — about 24" forward from the tail and close to the spine — then glue it in place.

10. PLANK THE TOP

It's time to put on the top skin. If you're planking on the frame as you go, keep on going (Figure **J**) — just remember to put in the vent box when you're ready. Again, you want to ensure that no twist is developing in the board. Place large weights on the top skin to iron out any warps or wrinkles, and prop the bottom with blocks, to alleviate the dreaded twist (Figure **K**).

When the glue is dry, clean up the skin by

carefully cutting it back to the outside edge of the stringers and the first and last ribs (Figure **L**).

11. BUILD THE NOSE BLOCK

The nose block is thick, about 1½" or so after shaping, 24"×3½"×2½" thick to start. If you have a big enough piece of wood to use, that's wonderful. Otherwise, clamp and glue up smaller scraps (cedar is great). The sensuous curves of the nose and tail pieces look amazing with laminated pieces of wood, so I always go that route.

Glue the nose block to the first rib on the bow end of the board.

12. LAY OUT THE RAILS

Now we're getting to the heart of my NeverSink method. If the board suffers a terrible crack, these rails will keep it afloat.

We're using 2"-thick extruded polystyrene (XPS) foam, commonly found as "pink" foam in home improvement stores. Wear a mask when you cut or burn this material. Cutting XPS is not that bad with a sharp hand saw. A hot-wire setup works great, though the fumes are pretty noxious. A table saw is the best way to cut it, however.

Rip four 4" wide strips of XPS, each 8' long. Then cut a 45° angle on one end of each rail (looking down on the 2" edge). This is where you'll scarf the rails together, amidships. Stand the rail pieces up against the rail stringers just for orientation, with their 4" faces vertical and their 45° cuts at the middle (Figure **M**). This is your basic layout.

You'll notice the bow portion of each rail has quite a bend to it. To alleviate this and ensure that the foam rails don't crack when we're torturing them to this curvature, we'll use *kerf cuts*. Using a sharp saw, knife, or hot wire, cut notches about 1" deep from outboard to inboard, about 2" apart, from the nose to about 30" back (Figure **N**). These kerfs allow the foam to bend without over-stressing and rupturing its skin. You can use this technique on wood as well!

Now you can glue the rails to the rail stringers.

13. BUILD THE TAIL BLOCK

Just as you crafted the nose block, now craft the tail block. You can place this on the board before

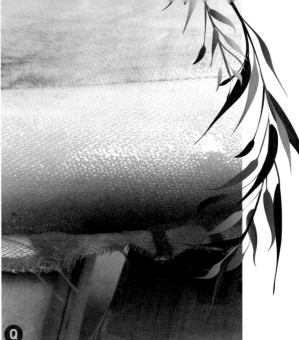

bringing the rails in, or after bending them around the stern-most rail curvature. Glue the block tightly to the last rib; you can run ropes or straps along the long axis of the board to cinch it tight.

14. SHAPE THE NOSE, RAILS, AND TAIL

Now you can cut down and shape the nose, rails, and tail to the profile of your choice using a fine handsaw, then the Surform rasp, a hand plane (Figure **O**), and/or varying grits of sandpaper. If you use a hot wire to shape the foam, beware of slicing away too much!

Perhaps the easiest and best rail profile is a D-shaped cross-section that has the same curvature on top and bottom ("50-50" rails). Of course you can try alternate profiles such as a "pinched" rail, but these are for specific performance aspects.

15. EPOXY THE BOARD

I won't go into all the wonderful traits of epoxy, or exhaust the many ways in which you can lay up your board. Suffice it to say, you need at least one coat of epoxy on the board to seal it. First, patch any gaps in the planking or rails. You can mix some fine sawdust with epoxy to create a patching compound the consistency of peanut butter.

To seal the board, apply the epoxy with a foam roller or carefully with a brush. Beware of drips, as these will cause sanding headaches later!

Optionally, you can put a layer of fiberglass on the top (deck) and/or bottom of the board for extra abrasion resistance, but you'll have to fully "wet out" the fiberglass cloth and then add at least two or three layers of epoxy to fill up the weave.

16. FIBERGLASS THE RAILS

The XPS foam rails must be covered with fiberglass for strength. First, you can fair the rails (spackle any gaps) with simple drywall joint compound. After fairing, coat the rails with epoxy and then lay the fiberglass cloth on the wet epoxy. Rails, nose, and tail blocks should all have a layer of 4oz cloth that overlaps the joint between rail and skin; I recommend glassing the entire deck while you're at it (Figure **P**). "Wet out" the cloth fully with epoxy, smooth out any bubbles, and clean up any drips.

When this first laminating coat is dry (Figure **Q**), you'll want to brush or roller on at least two more coats of epoxy on top of this, to completely fill the weave and create a smooth surface for sanding. Plan on a few days.

17. MAKE YOUR FIN

The surfboard fin or *skeg* is a simple creation that you can cut out with a jigsaw or your CNC. Typically, these are about 8" long for a board this size and there are many shapes; search the internet for "center fin" and you'll see the

R

S

T

U

dizzying variety. For a good all-around fin, a large "slanted D" style, which looks pretty fish-like, is fine (Figure **R**). Make this out of two pieces of ¼" plywood laminated together, and sand down the leading edge to a graceful, blunt curve. Sand down the trailing edge to a sharper profile, but don't make it knife-sharp, as that will actually hinder performance.

Cover the fin in a coat of epoxy, or optionally, fiberglass it too.

18. LAY OUT THE TAIL FIN
Epoxy the fin on the bottom of the board, about 8" forward from the tail (Figure **S**). For strength, create a fillet from epoxy and sawdust and apply it all around the board-to-fin joint with a gloved finger.

TIP: Come back in an hour or two and dip a gloved finger in denatured alcohol and then smooth the fillet joint; it does wonders and cuts down on sanding later.

19. PAINT THE RAILS
You're getting really close now! Since the rails are that unsightly pink foam and drywall compound, they won't look all that great. Paint them, and any other areas you wish, with a good all-purpose paint for plastics such as Krylon (Figure **T**). Go slowly and sand between coats to get it smooth.

20. FINISHING
Finish work — you love it or hate it. Here you'll sand, paint, sand, and sand some more. Make sure

you're not sanding through any fiberglass cloth! Add any extra designs, if you like.

21. VARNISH

As epoxy is inherently not UV-stable, you'll want to varnish over it. Pick a good spar varnish, go slowly, and avoid lumps and sags (Figure). You can varnish over the painted rails as well, though it may impart a slight color change. I sometimes add pinstripes and I varnish over these for extra protection; the coloring actually is quite pleasing.

SURF'S SUP!

You're done! Sure, there may be odds and ends, and sure, the overall curvature of the board may not be exactly perfect. But it's close enough, it's beautiful, and it's yours. You have come a long way and put your hand into many different techniques from CNC to artistic joinery and composite hull construction. Hooray!

Once your board is dry, you can create a paddle (I like the one at instructables.com/id/Build-a-Stand-Up-Paddleboard-Paddle) or just go out and hit the waves.

I recommend you don't let your SUP sit in the hot sun or a sealed-up vehicle for long, if you can help it. But the vent really works well and should mitigate any problems.

Keep your board clean and keep grit from grinding into the varnish coat. If it gets a lot of use, varnish it again every couple of years.

As always, think safety. Take a class or get instruction. Wear a life-preserving device and never go too far from a safe landfall. Watch the wind and weather, and have fun! ◗

For more photos and tips on materials, finishing, and workshop safety, visit the project page at makezine.com/go/cnc-sup.

SONNY LACEY builds watercraft in his incomprehensibly small Baltimore row-house basement-turned-workshop. He hopes to someday see boat-building return to Charm City.

Surf STAND

Assemble this **surfboard rack** to keep your quiver clean and clear

Written by Alexey Volochenko

Y ou've got your board, now you should store it safely. The Alexey Surfboard Rack is a free-standing rack that you can build to hold your boards vertically. It uses a rope floor and back to protect the fragile edges of your boards, and the open-source plans come in a variety of sizes to fit anything from 3 to 8 boards.

The wooden components for the rack are designed to be milled from birch plywood on a CNC router, making for a tight fitting design that is precise enough to incorporate captive nuts and dados. It also means that all the cuts are perfectly straight and consistent, resulting in a good looking outcome. ⊘

Download the design files at
obrary.com/collections/open-designs/
products/alexey-surfboard-rack

ALEXEY VOLOCHENKO
is a software engineer and a tech enthusiast with a passion for surfing

Traversing sand dunes and navigating bay waters, the 3-day, multi-terrain human-powered **Kinetic Race** celebrates 50 years with 50 miles of grueling and hilarious ingenuity **Written by Robert van de Walle**

Pedal TO THE MEDAL

One day in 1969, Jack Mays walked past his friend Hobart Brown's northern California studio and saw Hobart butchering his son's Justin's tricycle.

"What are you making?" Jack asked. "Whatever it is, I bet I can beat it in a race."

Word got out and a total of 11 entries competed in that first ever sculpture race on Mother's Day in Ferndale. Teams focused on capturing the hearts of the spectators rather than winning. Neither Hobart's *Pentacycle*, made from Justin's tricycle, nor May's entry, *The Tank*, finished. The crowd favorite and therefore judged to be the winner was Bob Brown's smoke-belching, egg-laying *Tortoise*.

Today, five decades of highly creative people have left their imprint on the Kinetic Race. They elect a Rutabaga Queen at a massive fundraising ball a week before the race. During the race The Great Razooly is blamed for all malfunctions, breakdowns, and chaos. A non-profit organization founded by volunteers runs the race. HAM radio operators alongside the course report racer positions. Local radio station KHUM devotes the entire weekend to covering the race. Businesses sponsor teams, and the local economy gets a boost of over a million dollars.

Some teams build their sculpture over the course of years; others start the day before the race. There's no "right" way to build. The only regulation is that stored energy such as from a pressure vessel or a battery can not be used to move the vehicle forward.

This is a race where the geniuses from Jet Propulsion Laboratory couldn't beat two guys on bikes with trash cans as their flotation. This is a race where when a team breaks down another team will stop and help them get back in the race. This is a race where the biggest cash value prize is for the team that finishes smack in the middle, winning the "Mediocre Award." This is a race for polymaths and brilliant team leaders. This is a race that understands what the external world values and turns it sideways.

Celebrating its 50th Anniversary this past Memorial Day weekend, the race continues to inspire engineers and artists, spectators and athletes, young and old, rigid and free spirits, rule followers and rule breakers. It's not a wacky race. It's a human race. ⊘

Learn more and get involved:
kineticgrandchampionship.com

ROBERT VAN DE WALLE is an artist, inventor, fabricator, athlete, and explorer. He uses mind and muscles to do fun stuff.

#VANLIFE

Why be constrained to just one campsite? Convert a van, pack up your gear, and the whole world becomes yours to explore. Written by Caleb Kraft

Van camping — or even living in one long-term — has made a recent resurgence, spawning hashtags like #vanlife across the internet and putting thousands of people on the road in their wheel-laden micro-homes.

These campers can range from stealthy conversions, intended to blend into any city street or parking lot, to monstrous cabins on wheels, designed to take on trails where no vehicle has tread before. It's nearly as exciting to imagine trekking in all the variations as it is to actually do so; here are a few builds that stand out to us as particularly inspiring.

EARTHROAMER HD
earthroamer.com/hd

This extreme beast supplies luxury living on top of a Ford F-750 4×4 frame. There's not much that will stop you from getting to your target destination, and once you get there the massive stores of fuel, water, batteries, and solar panels will make sure that you have a pleasant stay.

Earthroamer, Marina Piro, faroutride.com, Ryan Willis

SAVINGS VOUCHER

MAKE Magazine	COVER PRICE	YOU SAVE	PAY ONLY
One Year	$59.96	33%	$39.99

Get it now!

☐ PAYMENT ENCLOSED
☐ BILL ME LATER

MAKEZINE.COM/SAVE33

B99NS2

NAME _____ (please print)

ADDRESS/APT. _____

CITY/STATE/ZIP _____

COUNTRY _____

EMAIL ADDRESS (required for order confirmation) _____

Make: currently publishes 4 issues annually. Allow 4-6 weeks for delivery of your first issue. For Canada, add $9 US funds only. For orders outside the US and Canada, add $15 US funds only.

CALEB KRAFT has an obsession with vans and busses. He has owned six vintage vehicles of the boxiest persuasions and loved them all immensely.

PAM THE VAN

pamthevan.com

Stealth is the name of the game here. The van, named Pam, could park anywhere and you'd have no idea that its builder's home was inside. Marina Piro built out the interior herself, constructing bed and storage structures to fit her needs exactly, and she really seems to get good use out of it, traveling the world with her dog Odie.

FAROUTRIDE

faroutride.com/van-tour

This is a customized modern high roof van, a common choice for people looking to build their own tiny home on wheels. A knotty pine interior makes it feel more like a Tahoe cabin than a Ford Transit. Its owners sold their house and quit their jobs to convert this and go on an open-ended road trip, all chronicled meticulously on their site.

VW MICROBUS

kombilife.com

We can't talk about camper vans without mentioning the iconic vehicle that started the trend, the Volkswagen bus. With the fresh popularity of camping on the go, the price for these old beasts has gone up, but if you look for a model from more recent years, you may be surprised at how approachable the cost can be!

Backyard
PUMP TRACK
Design and build your own mountain bike
track for riders of all skill levels Written by Alex Paveles

TIME REQUIRED:
A Weekend

DIFFICULTY:
Intermediate

COST:
Free–$10,000

MATERIALS
- » **Dirt** You'll probably want to import dirt rather than quarry it from your yard. *Fill dirt* can be free; *finish dirt* prices vary depending on the mixture. A standard pump track might need 150–200 cubic yards of dirt, but digging one in your yard takes way less.
- » **Water**
- » **Concrete mix or clay (optional)** to mix into your soil if it's sandy. Pure clay will crack like crazy. A mixture of *DG* (decomposed granite) and clay makes for good dirt.

TOOLS
- » **Shovels**
- » **Rakes**
- » **Special digging tools (optional)** such as a grub hoe, digging/tamping bar, etc. The *Pulaski* is a combination grub hoe and adze/axe, while the *McLeod* is a combination wide hoe and rake. Both were developed by forest firefighters.
- » **Axe, saw, loppers (optional)** for clearing tree branches and roots
- » **Bobcat loader (optional)** great for bigger tracks

Ⓐ

Ⓑ

Mountain bike pump tracks are great for riders of all ages and skill levels. They provide a place for kids to learn how to *pump* — to propel the bike using your body motion instead of pedaling — and how to corner a bicycle and hit jumps. And they're simple to make if you've got the space. Thanks to DIY pump tracks, even advanced and pro riders can perfect their techniques with the convenience of riding at home.

Professional mountain biker Mark Weir was an early adopter, building a pump track in his backyard in 2005. Word spread quickly in the mountain bike community and riders started building their own pump tracks. What the backyard mini ramp is to a skateboarder, the pump track is to a mountain biker. We're not just riders, we're *diggers*.

You don't have to be an expert to build your own backyard pump track — you just need to be willing to pick up a shovel and start digging. For a bigger yard, it's nice to have a small Bobcat to get lots of dirt moved around and get the main piles going. Then you shape with shovels and rake, and get the dirt to a certain moisture so it packs. The sun makes it hard, and so does riding it.

1. DESIGN AND DIG
There's no one track design that works for every spot; you have to build and design to fit the landscape you have. I start with the outer lines and set where the corners will be, and then fill in the middle as I go.

1a. Having a high spot for a *roll-in ramp* is key, to generate initial speed without pedaling (Figure **Ⓐ**).

1b. Next you'll want to place *berms* — these are the banked corners and turns of your track (Figure **Ⓑ**). You don't want too tight or too mellow of a radius, but if you have a small backyard the berms may end up being tighter than you'd make them in a larger space.

1c. Now start adding the *rollers* (Figure **Ⓒ** on the following page). The top of a roller shouldn't come to a peak, but should be rounded. A mellower roller will allow you to generate more speed than a steep, peaky roller.

The roller spacing that works best for most

POST OFFICE TO PUMP PARKS

With the popularity of backyard pump tracks growing, it didn't take long for riders to encourage their cities to build bike parks. Aptos, California, was one of the first. Dubbed **Post Office** for its close proximity to the Aptos postal station, this small temporary park slowly became one of the most well-known jump spots in the mountain bike world. Built by volunteers, and blessed by the private landowner and the local parks department, Post Office (1986–2015) was a great example for other cities to see how important it was to have public riding spots, and it became a model for developing **community pump tracks**.

Nowadays, pump tracks are in just about every town and city that has a skate park. The newest development is **asphalt pump tracks**, led by Velosolutions, a company out of Switzerland. The use of asphalt tracks is a great innovation for public bike parks because there's little to no maintenance involved. Starting with 2018, Velosolutions and Red Bull now sponsor an annual **Pump Track World Championship**.

layouts is having one bike length in the flat bottom between two rollers. As you pile up the rollers, you'll want to shape the bottoms to continue the same radius as the rollers. This will help you keep your flow and speed going when you're riding it.

There are endless combinations of rollers you can add into your track. Get creative and try out different lengths and layouts to fit your riding style and the space you have to work with.

1d. Once you have an outer loop that allows you to pump laps, you can start adding *transfer lines*, *U-turns* (Figure **D**) and other ways to change direction on the track. Test-riding while you build is key to make sure everything lines up properly. That way, you don't have to backtrack on your build.

2. PACK IT IN

There are many ways to pack your track in. You can slap-pack it with flat-head shovels (Figure **E**) or use a Vespa scooter, mini moto (pocket bike), power rammer or plate compactor, golf cart, or quad with slick tires, and of course your bike.

Getting the dirt wet but not muddy and mixing it well is key for the dirt to pack and stay hard. Every type of dirt needs a different amount of water. If you have sandy dirt you can mix concrete or clay into it to make it stay hard and not fall apart every time you ride it.

3. ADD FEATURES (OPTIONAL)

Jumps are a favorite feature (Figure **F**). If you have room, you can add lines with *gap jumps* (aka double jumps) for advanced riders, so beginners can go around them.

Adding wood features to create *wall rides* is a fun and simple way to take your track to the next level. A good example is the College Cyclery pump park near Sacramento, designed and built by Randy Spangler (Figure **G**). He took the concept of a skate park and created the ultimate dirt-and-wood pump park for bikes. There's now a pump competition at the annual Delta Dirt Jam in September, and the pump park is open to the public to ride during the event weekend.

KEEP IT ROLLING

A dirt pump track needs upkeep in order to keep it

C

D

E

F

G

running smooth and fast. The best thing you can do is *water and sweep* with a push broom each time before you ride.

This is easy if it's a backyard track, but can be more difficult at a city park. It's very beneficial to work with your parks department to establish a maintenance crew — not only does it keep the track running, but it also gives mountain bike enthusiasts and diggers the opportunity to establish careers for themselves.

Your pump track is never fully finished — you can always add more features to it once you've mastered riding it and are ready to progress your riding. ◉

More on building pump tracks:
mbaction.com/how-to-build-a-mountain-bike-pump-track-story-video

Set up your bike for pumping:
ridemorebikes.com/best-pump-track-bike-setup

ALEX REVELES was born and raised in Santa Cruz, California, and grew up riding mountain bikes at the Aptos Post Office jumps. His love of digging and riding dirt jumps led him to building pump tracks and digging at the Red Bull Rampage.

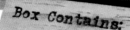

Box Contains:

1 Kathy SCREECHER

& 1 NeRVe ShaKeR

Make a Sweet
ALUMINUM WHISTLE

Impressive yet easier than it looks, it's perfect for outdoor adventures or gift giving

Written and photographed by Nathaniel Bell

It's a well-known fact that everyone should own at least two good whistles. In an emergency on the trail or on the water, a whistle can call for help much louder — and farther —than your voice can. So if you're reading this and you have less than two, you better run out to the nearest home store and pick up an aluminum rod and tube.

I've made a few whistles over the years out of various materials. I'm not sure why, but a smooth, shiny, handmade aluminum whistle impresses people way more than a wood whistle. Aluminum is a pretty easy material to work with — most woodworking tools will work fine to shape and sand aluminum too.

1. CUT THE TUBE

Cut a piece of your tube stock about 4"–5" long. I usually cut it long and trim down later. The final length will affect the pitch, but you'll tweak that in a later step. You can use round or square tubing. I'm choosing round (Figure **A**); I think it feels a bit more comfortable on the lip.

2. FILE THE WINDOW

If you have one, a triangle-shaped file is perfect for cutting the *window* — the opening in the top of the whistle. Start filing a groove ½" to ¾" from one end of the tube; this will be the mouthpiece end (Figure **B**). Keep the groove straight on the mouth side, but angle it on the opposite side to form a wedge shape as shown in Figure **C** — this edge is the *lip* or *blade* the air will be blown against, creating the whistle sound. Getting this edge right is where half of the whistle's magic comes from.

TIME REQUIRED:
1–2 Hours

DIFFICULTY:
Easy

COST:
$10–$20

MATERIALS
» **Aluminum tube stock, round or square**
» **Aluminum rod stock** that matches the interior dimensions of the tube stock
» **Epoxy (optional)** or brazing supplies, if rod stock doesn't fit tightly inside tube stock

TOOLS
» **Hacksaw or bandsaw**
» **File** A triangle file is ideal but other files will work too.
» **X-Acto knife**
» **Ruler or measuring tape**
» **Hammer**
» **Sandpaper**
» **Steel wool**

OPTIONAL BUT NICE TO HAVE:
» **Vise**
» **Drill or drill press**

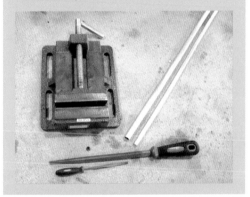

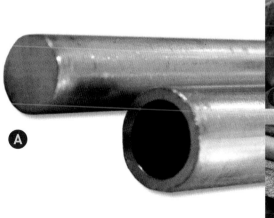

Use an X-Acto knife to clean up the inside of the window and to remove any burrs (Figure **D**).

3. CUT THE FIPPLE PLUG

Measure from the window to the mouthpiece end of the tube, and cut a piece of the rod the same length. This plug is sometimes called the *fipple*.

File or sand one side flat, just a hair (Figure **E**), to create the *windway* — the thin passage you'll blow air through. I find giving it a bit of a taper back toward the mouth side helps with the overall sound later on.

4. TAP IT IN

Line up the flat side of the fipple plug with the top of your whistle and hammer it into the mouthpiece end. Be sure to keep the two lined up as you tap it into place (Figures **F** and **G**). If you need to adjust the alignment, I find sticking a pencil in the open end and hitting it against a table will usually do the trick and dislodge the plug.

Once you have the mouthpiece in and aligned, by plugging the other end with your finger, you kind of sort of have a working whistle. If you're a slacker, you can just stop here and say you made a whistle. But you're not — let's finish off this potential headache creator.

5. TEST THE SOUND

Place your finger over the end and give it a blow. If you only hear "air," then your window is probably too small. Using your file, adjust your window and lip until you get a nice eardrum-splitting pitch.

6. TUNE THE PITCH

This is where the other half of a whistle's magic comes from: The length of the whistle's inner chamber also affects the overall pitch, and you can change the chamber's length by using different sizes of plug in the other end of your whistle.

Using a pencil (or the rod stock, if it moves around inside easily enough) as a temporary end plug, move it up and down while blowing into the whistle to find your desired pitch. You'll notice a "slide whistle" effect — but slide whistles are for kids, and we're making a serious adult whistle with a single pitch! A longer chamber will have a warmer "train whistle" sound; a short chamber will have a sharper, higher-pitched sound.

For my whistle, I found that plugging up the bottom 1" hit the pitch I wanted. Cut a plug to your desired length (Figure **H**).

7. FIT IT ALL TOGETHER
My rod stock fight tightly inside my tube stock and required hammering (Figure **I**). If yours is loose, you'll need to epoxy or braze the pieces together.

8. TRIM THE END
At this point, you can cut your whistle to any length as long as the end plug goes in further than the place where you're cutting (Figure **J**). My inner chamber is only about 1" long, but I like keeping the outside length longer so it's easier to handle.

9. SAND BOTH ENDS
Using a belt sander or your files, sand the ends flat. If your rod stock is tight fitting, you should be able to get the appearance of a single solid piece with just sanding. Making the whistle look like it's made from a single block of material is sure to impress your coworkers (however, they'll most likely stop being impressed once they get sick of you continually blowing that whistle you made).

You probably want to give the ends a bit of a chamfer so that you don't cut up those beautiful lips of yours every time you use your whistle (Figure **K**).

10. POLISH YOUR WHISTLE
Polishing can be done by hand, but if you're using round stock, a drill or a drill press sure will come in handy here. Wrap the end of the whistle that will be in the chuck with either painter's tape or a piece of paper to prevent it from getting marred by the inside of the chuck (Figure **L**). Turn the drill on a low speed and sand and buff away. I usually start off with some high-grit sandpaper to sand out any scratches or dings and finish up with steel wool (Figure **M**).

That's it! Give it a name and give it a blow!

NATHANIEL BELL is lead artist at Insomniac Games NC, a father of three, chicken aficionado, and maker of things in Hillsborough, North Carolina. He wrote "Light Switch Complicator" in *Make:* Volume 67 (makezine.com/projects/light-switch-complicator). Follow him at instagram.com/stuffnatemakes.

BE PREPARED

Get the most out of your camping trip with these clever projects

Written by Caleb Kraft

There are few moments in life as relaxing as lying under the stars, listening to the crickets chirp. A lone cricket can be torture, but the chorus of nightfall while camping is glorious.

As any industrious maker knows, half the fun of camping is in the preparations you take before you go: the gathering of supplies and most importantly, the construction of the items you'll need on your adventure.

Here are a few camping projects that will allow you to take your wilderness fate into your own hands, and enjoy the satisfaction of using gear you've made yourself.

CALEB KRAFT has spent more than a handful of nights under the stars, and even more dreaming of it.

TIN CAN COOKER

It's arguably one of the most common and most fun projects for camping: Take a worthless old tin can and turn it into an efficient cookstove. If you've cooked a meal over a raging campfire you'll know it's a hot, messy job that often leads to burnt supper. Fueled by a tuna can full of wax and cardboard, **Kathy Ceceri's** tin can stove gives you the control to cook a nice meal in comfort while you watch the bonfire from a distance. makezine.com/projects/tin-can-cooker

CAMPING HAMMOCK

A tent may pop into your mind when you think of camping, but they're often difficult to assemble, they're hot and muggy, and they block your view of the stars. Why not sew **Andreas Kullerkann's** camping hammock instead? Its tension string lets you sleep nearly flat, and during the day it's the perfect platform for lounging. instructables.com/id/Camping-Hammock

SOLAR POWERED BATTERY CHARGER

At some point, when you're absolutely sure you've become one with nature, you're going to remember that Instagram exists. How are you going to keep that phone battery topped up and your feed filled with vibrant images of how relaxed you are? With **Forest M. Mims III's** smart solar charging circuit. makezine.com/projects/solar-power-battery-charger

SEWABLE SILVERWARE HOLDER

If you camp regularly it's nice to have a utensil set that's ready to go, packed and clean. Take an afternoon to sew together **Susie Osborn's** decorative roll-up silverware holder and it'll keep your eating tools from getting mucked up with your other gear. makezine.com/projects/diy-camping-silverware-holder

ULTRA SIMPLE CAMPING TOILET

When you're camping in wilderness, "leave no trace" means you pack out everything you packed in, including your poo. Squatting over a plastic bag is actually more difficult than it may sound, so **Kevin Carruthers'** addition of a simple bucket with a little luxury padding makes a world of a difference. instructables.com/id/Camping-Toilet

Get Nybbled!

Build a true quadruped, walking robot kitten with the OpenCat framework running on an inexpensive Arduino

Written and photographed by Rongzhong Li

TIME REQUIRED:
1-3 Days

DIFFICULTY:
Advanced

COST:
$150–$300

MATERIALS
» **Petoi Nybble quadruped kitten robot kit**
$225 at petoi.com, includes:
  » **NyBoard V0 custom microcontroller board** Arduino compatible, with IR receiver
  » **Laser-cut wooden body frame parts**
  » **Servomotors, metal gear, high torque (11)** with cables and servo horns
  » **Ultrasonic distance sensor**
  » **Springs (11)**
  » **Infrared remote control**
  » **FTDI to USB converter** for uploading code to NyBoard
  » **LED and resistor**
  » **Battery holder with on-off switch, 2×AA/14500 size**
  » **Various screws**
» **Li-ion batteries, 14500 size, 3.7V (2)**
Don't mix with regular AA batteries (1.5V)! I've been using EBL 800mAh batteries and they work well.
» **Smart charger** for Li-ion batteries

OR:
» **DIY OpenCat Mini** You can build your kitten robot from scratch using similar electronics sourced separately, a 3D-printed frame, and cheaper, plastic gear servos. The hardware and software are well documented at create.arduino.cc/projecthub/petoi/opencat-845129, github.com/borntoleave/catMini, and petoi.com/forum.

OPTIONAL:
» **Raspberry Pi single-board computer (optional)** if you want to add AI; not needed for preprogrammed gaits and behaviors.
» **HC-05 Bluetooth module** to wirelessly upload code and communicate
» **Paints and 3D printed accessories** to give your Nybble a unique look

TOOLS
» **Utility knife**
» **Screwdrivers, slotted and Phillips**
» **Computer with Arduino IDE** Install the latest Arduino IDE
» **Cable, USB to mini-USB** to connect the uploader to computer. Not micro-USB.

OPTIONAL:
» **Soldering iron and solder** to solder the decorative LED to ultrasound sensor
» **Multimeter** to test and debug
» **Oscilloscope** to test and debug
» **Hot glue or super glue** I avoid using them. OpenCat is designed to be soft!

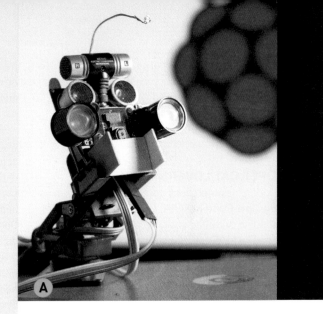

(A)

Three summers ago, I had two graduate degrees from Wake Forest University in North Carolina, but still didn't know where to go. I bought an inexpensive Raspberry Pi single-board computer just to test some of my previous codes.

While setting up the camera and other peripherals on the Pi, I realized its huge potential. I liked to do handcrafts and hands-on science experiments, but my academic training was mostly theoretical. I kept a list of creative ideas, and the Pi became the key to an arena where all my skills could merge to realize those ideas.

I always wanted to make a robotic hand or a walking animal like Boston Dynamics' famous quadruped robots. The pan-tilt camera on my Pi looked like a cat's curious eye (Figure (A)). Why not set it free on its own feet?

FREE FROM THE BURDEN OF KNOWLEDGE

Because there were very few quadruped robots being built with cheap servos, I didn't have the burden of knowledge to limit my imagination. Skills and intuition from my previous research in biophysics and computer science helped a lot in designing the cat. I used hobby servos and Popsicle sticks to build its body, and wrote the Python and Arduino code from scratch. Every time I learned a new hardware module, I would find a place to apply it on the cat.

In one month, the cat was able to walk around and do tricks like obstacle detection, trotting, rolling over and standing up again, and dynamic rebalancing after disturbance, just like the famous BigDog videos. After one year, the cat grew into a 3D printed skeleton hosting many cool features: face tracking, waving hello in greeting, touch detection, sound effects, even Alexa integration (Figure **B**).

OPENCAT LOVERS ONLY

During these iterations, I abstracted the design as the OpenCat framework, combining Raspberry Pi, Arduino, and other easily accessible components. I had no funding or investment, so it had to be cost-effective. And I wanted it to be a generic

framework for walking robots, so it should not be complex and expensive. Of course, simple and cheap solutions usually come from complex derivations and expensive trials and errors.

For a foreigner in the United States, working on an "unauthorized" personal project instead of permanent employment meant a countdown of my stay for the American Dream. However, as the cat kept evolving, it became more demanding and dominated my heart. I released the first OpenCat demo on Maker Share in 2016 and later on Hackster and other platforms to see people's feedback. Thanks to the warm encouragement over the internet, I finally decided to follow my heart. I founded Petoi as the company to support OpenCat. My last year in the U.S. was dedicated to kickstarting Petoi, then I returned to China for production.

I realized there's a gap between makers and mainstream society — between their creations and the public's need. It's fine to be a hobbyist maker for a small community, but what if the project is too complex and expensive? What if we want to present it to the public and let more people enjoy our creation? It was time to go out and negotiate with a bigger world. I moved to Pittsburgh and

B

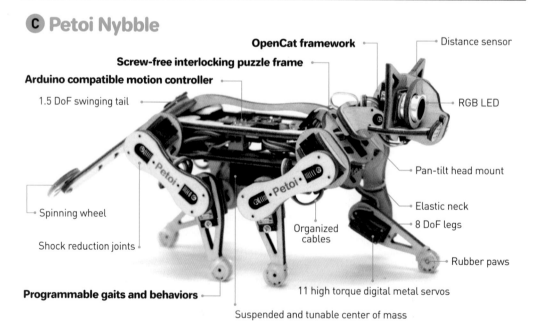

C Petoi Nybble

OpenCat framework

Distance sensor

Screw-free interlocking puzzle frame

Arduino compatible motion controller

1.5 DoF swinging tail

RGB LED

Spinning wheel

Pan-tilt head mount

Shock reduction joints

Elastic neck

8 DoF legs

Organized cables

Rubber paws

Programmable gaits and behaviors

11 high torque digital metal servos

Suspended and tunable center of mass

D Screw-free Interlocking Puzzle Frame

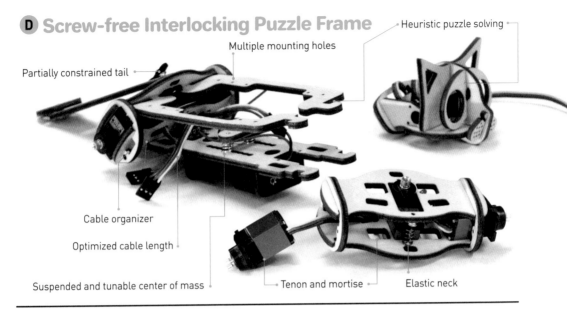

Heuristic puzzle solving

Multiple mounting holes

Partially constrained tail

Cable organizer

Optimized cable length

Suspended and tunable center of mass

Tenon and mortise

Elastic neck

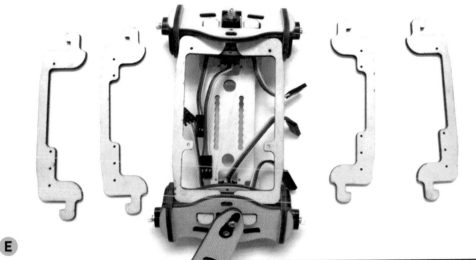

E

discovered Factory Unlocked, where I worked part-time in return for membership to develop the product and do small-batch production. I found collaborators from Pittsburgh, Stanford, Wake Forest, and China to help me polish the design and launch an Indiegogo crowdfunding campaign.

A KITTEN IS BORN

The original OpenCat was too hard to reproduce, so I designed Nybble as the first product (Figure **C**). Nybble's smaller, kitten-size body has a retro wood frame (in honor of its Popsicle-stick ancestor) that's assembled like an interlocking

puzzle, inspired by traditional Chinese woodwork (Figures **D** and **E**). It uses springs for elastic connections in the thighs and neck, to absorb shock, create additional degrees of freedom, and extend the life of the servos. Cats are soft, like liquid — the Taoism in this design is to adapt to disturbance, not to fight it. The battery holder doubles as a suspended, tunable center of mass.

Nybble is probably the lightest and fastest robotic cat in the world that really walks. It stores instinctive "muscle memory" to move around, so you can program in your favorite language, and direct Nybble to walk around simply by sending

F NyBoard V0

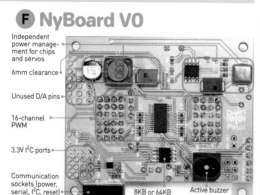

Independent power management for chips and servos

6mm clearance

Unused D/A pins

16-channel PWM

3.3V I²C ports

Communication sockets (power, serial, I²C, reset)

8KB or 64KB external EEPROM

Active buzzer

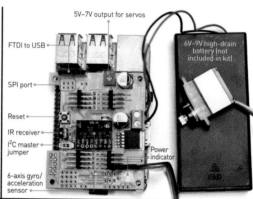

5V–7V output for servos

FTDI to USB

SPI port

Reset

IR receiver

I²C master jumper

6-axis gyro/ acceleration sensor

6V–9V high-drain battery (not included in kit)

Power indicator

ATmega328P (5V, 20MHz)

Raspberry Pi 3B+ (not included in kit)

Communication

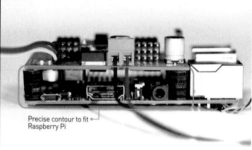

Precise contour to fit Raspberry Pi

G

short commands, such as "walk" or "turn left." The core motion code was extracted and transferred onto our customized Arduino board, NyBoard (Figure F), which used the same chip as a regular Arduino Uno. A lot of algorithmic optimization was made to utilize every byte available. (*Nibble* means a small bite from an animal, as well as half a byte.)

A Raspberry Pi can still be mounted on top to give Nybble an AI mind to help with perception and decisions (Figures G and H), but it's not needed to make Nybble walk — users only need to assemble the puzzle frames, connect the electronic components, then upload the Arduino code and calibrate the servos. The code is all published on GitHub (github.com/PetoiCamp/OpenCat) and I've shared assembly instructions and a tutorial playlist at Petoi.com.

We announced our launch of Nybble only through social media (@PetoiCamp). Thanks to sharing by maker communities, the demo video went viral and reached an audience of millions. Nybble was mentioned by top tech news outlets such as *Make:*, *TechCrunch*, and *IEEE Spectrum*, and our kitten robot appeared on TV to entertain even more people.

H

GIVE A GIFT
ONE YEAR ONLY $39.99
Make:

GIFT FROM

NAME _____ (please print)

ADDRESS/APT. _____

CITY/STATE/ ZIP _____

COUNTRY _____

EMAIL ADDRESS (required for order confirmation) _____

☐ Please send me my own subscription of Make: 1 year for $39.99.

We'll send a card announcing your gift. Make: currently publishes 4 issues annually. Allow 4-6 weeks for delivery of your first issue. For Canada, add $9 US funds only. For orders outside the US and Canada, add $15 US funds only.

GIFT TO

NAME _____ (please print)

ADDRESS/APT. _____

CITY/STATE/ ZIP _____

COUNTRY _____

EMAIL ADDRESS _____

499GS1

❶ OpenCat Framework

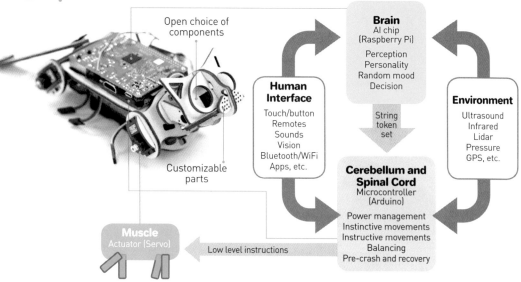

Open choice of components

Customizable parts

Brain
AI chip
(Raspberry Pi)

Perception
Personality
Random mood
Decision

Human Interface
Touch/button
Remotes
Sounds
Vision
Bluetooth/WiFi
Apps, etc.

String token set

Environment
Ultrasound
Infrared
Lidar
Pressure
GPS, etc.

Cerebellum and Spinal Cord
Microcontroller
(Arduino)

Power management
Instinctive movements
Instructive movements
Balancing
Pre-crash and recovery

Muscle
Actuator (Servo)

Low level instructions

BUILDING YOUR OPEN SOURCE QUADRUPED

Here's an overview of the OpenCat framework for quadruped robots, and the Nybble kitten kit. Legged motion is challenging in both software and hardware — now you have a cheaper platform besides simulation to study dynamic balancing and motion planning. Later when you're ready for more precise and powerful actuators, you can scale Nybble up to a dog, a horse, or even an elephant!

OPENCAT FRAMEWORK

During the 2.5 years' evolution of OpenCat, a rough framework became clear (Figure ❶). It started from a Raspberry Pi kit, introduced an Arduino as the central pattern generator, and is open to extensions and upgrades. If the diagram appears confusing, just think about how you walk without thinking, and how you plan a jump across a gap.

The OpenCat motion control circuit (Figure ❶) is a quite conventional master-slave structure,

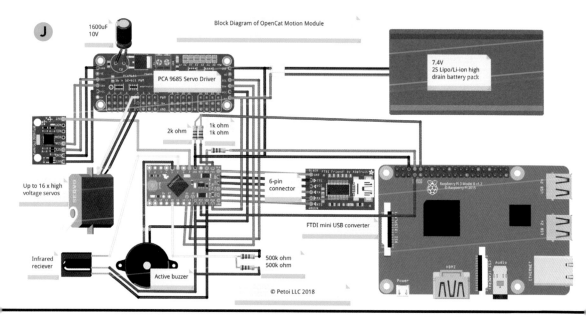

Block Diagram of OpenCat Motion Module

1600uF 10V

PCA 9685 Servo Driver

7.4V 2S Lipo/Li-ion high drain battery pack

2k ohm

1k ohm 1k ohm

6-pin connector

Up to 16 x high voltage servos

FTDI mini USB converter

Infrared reciever

Active buzzer

500k ohm 500k ohm

© Petoi LLC 2018

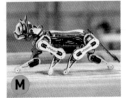

and this block diagram works for a DIY version of Nybble. All the adaptive movements you see are achieved on an ATmega328P chip, the same one on the Arduino Uno. Of course there are faster chips at quite affordable prices. I'm just challenging myself on the question "How much information does a machine need to walk?" By limiting the power of the motion controller, it helps to clarify the duties of the hierarchical control layers. People with different expertise can contribute and optimize their part. It also helps to keep the rhythm of walking without disturbance from other threads.

Makers in the OpenCat community have contributed code, 3D printed body frames (thingiverse.com/thing:3384371), a Bluetooth Arduino gamepad controller (thingiverse.com/thing:3624838), an Android app (thingiverse.com/asset:173883), and more.

NYBBLE AND NYBOARD

NyBoard is designed for two use cases: for Nybble's metal servos, and for DIY robots that may use plastic servos. Plastic servos can only stand 6V so there is a step-down chip on NyBoard.

Even without a Raspberry Pi in the kit, the code on Nybble is always listening for higher-level commands. You can send those commands either through Arduino IDE, Bluetooth UART, or serial connection to Raspberry Pi and other chips.

For example, if you want to move Nybble's leg, just type or encode the string **m8 30** and it will rotate Nybble's **8**th joint to **30** degrees. Or try **kwkL** — that means call Nybble's s**k**ill "**w**alk **L**eft" — and Nybble will walk left, without you controlling every detailed limb movement.

PROGRAMMABLE GAITS AND BEHAVIORS

After assembling Nybble, entry-level computer and Arduino skills are needed to upload the code and tune Nybble to working order. Then you can control a wide range of built-in behaviors and gaits from the infrared remote: stand balanced, and rebalance when disturbed (Figure **K**); walk and turn (Figure **L**); back and turn; trot (Figure **M**); crawl and turn (Figure **N**); wave hello (Figure **O**), crouch (Figure **P**), "butt up" (Figure **Q**), stretch (Figure **R**), and even "pee." More skills are stored on the board, such as push-ups, turtle roll, and jump; these may need more tuning on your robot and must be called from a serial terminal.

You'll need higher-level skills (or the willingness to learn) to enjoy all the tech behind Nybble and to teach it new tricks like high five (Figure **S**), sit up (Figure **T**), stand on hind legs (Figure **U**), or even a headstand (Figure **V**). These can be composited by transiting between a series of static postures (Figure **W**). I used a lot of high school math!

WHAT'S NEW, OPENCAT?

We shipped all the Indiegogo pre-orders before I left the U.S., and as people have received their Nybbles, a lot of fun discussion is going on at petoi. com/forum. Skilled makers are also designing their own OpenCat robots based on my old DIYable version (github.com/borntoleave/catMini), proving its potential as a generic walking platform.

Nybble will be compatible with future models of OpenCat, and we're developing new prototypes to meet more people's needs (Figure **X**). I just returned to China to scale up production — I'm renting a space at Seeed Studio's x.factory — and I hope to resume R&D as soon as possible.

Looking back on my journey, there were several hard transitions. For me, it was from an academic theorist to a hands-on maker, then to an entrepreneur. For OpenCat, it was from a personal project to teamwork, and then to a community project (Figure **Y**). For Petoi, it was from an idea in the U.S. to production in China.

By working on OpenCat, I wanted to show the potential of limited resources and how a maker's creation could be recognized by the public. By funding Petoi, I wanted to explore how an idealistic project could survive in this business world by generating real values. Success or failure, our story should be helpful to future makers who have great ideas. ◉

S

T

U

V

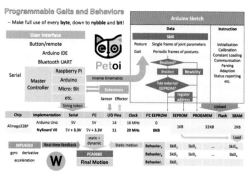
W

X

Y

RONGZHONG LI (RZ) is a physics Ph.D., computer science master, photographer, poet, and interdisciplinary explorer. He was an educator at Wake Forest University before becoming a full-time maker.

Follow this project at @PetoiCamp and igg.me/at/nybble, and see it at Maker Faire Shenzhen, November 9–10, 2019, shenzhenmakerfaire.com.

Makey Is Real!

Written by Camilo Parra Palacio

Build our biped mascot with all-new articulated arms and head, plus all the dance moves of the popular Otto platform

Yes, I've designed the first dancing Makey, the Maker Faire robot! Based on the same software and mechanical principles of the popular Otto biped robot (ottodiy.com), this new Makey can walk and dance, plus I've added new articulated arms and head so you can give your robot fresh moves. And you can build it in a weekend.

INSPIRATION

Maker Faire is the greatest show-and-tell on Earth! Ultimately this project is a gratification, a tribute to Maker Media, now resurrected as Make: Community so that thousands of people can continue to replicate and share Maker Faire events in their own cities.

From my experience making interactive robots, I thought it would be an interesting exercise to see how to "Ottomize" almost any robot character — to bring it to life using the same mechanical principles of Otto, our popular DIY biped robot (wikifactory.com/+OttoDIY). Otto is descended from the original BoB the Biped and Zowi robots (see *Make:* Volume 61, "DIY Bipedal Robot") and we recently evolved a new Otto "humanoid" with articulated arms.

For my first Ottomization, I chose the well-known Makey robot character from the Maker Faires. I selected Makey because it looks pretty similar to the structure we use in Otto, and although the articulations might vary, I thought it would be fun to make it move the way Otto does. I knew it could work.

I searched online and didn't find anyone who had taken on this challenge at such a small scale. I saw some amazing big (and mega) Makey designs, but no one had made an automatic walking biped robot of it. So I did.

TIME REQUIRED:
1–2 Weekends

DIFFICULTY:
Intermediate

COST:
$70–$99

MATERIALS

You can get the materials on this list in the Otto DIY Maker Kit M, or Builder Kit M (also includes the 3D printed parts), at ottodiy.com/#our-kits; a portion of proceeds benefits Make: Community. Or you can source them separately.

» **Servomotors: SG90 or MG90, 180° (5) and SG90, 360° (2)** We use metal gear servos for the legs, for strength.
» **Arduino Nano microcontroller, ATmega328 version**
» **I/O shield for Arduino Nano** such as Amazon #B01J4T7EQK
» **USB-A to Mini-USB cable**
» **Piezo passive buzzer, 5V**
» **LEDs (2)**
» **Resistors (2)** to limit current and protect LEDs
» **DuPont jumper wires, female to female (10)**
» **AA alkaline batteries, 1.5V (4) with battery case** for non-rechargeable option
» **LiPo battery, 7.4V** for rechargeable option
» **Micro switch, locking on-off button, 8mm×8mm, 6-pin** Amazon #B07FW48Z7H
» **3D printed Makey body parts** Download the free STL files from wikifactory.com/+OttoDIY/makey. You'll need a head, legs (2), feet (2), body bottom, and body top printed in red, plus the M logo and chest band in white.
» **Sensors (optional)** Makey is compatible with a wide variety of 5V sensors.
» **Bluetooth module (optional)** such as HC-05, HC-06, or BT-06, to provide remote control from a free mobile app.

TOOLS

» **3D printer (optional)** You can 3D print the parts yourself or buy them in a kit from Otto DIY (see above), or just send the files out to a service for printing.
» **Phillips screwdriver**
» **Soldering iron and solder**
» **Computer with Arduino IDE software** free download at arduino.cc/download

CAMILO PARRA PALACIO is a product designer, engineer, and founder of OttoDIY, a 3D-printable, open source robot than anyone can build, making STEAM education and robotics accessible to all. He's based in Olomouc, Czech Republic.

A

B Print-in-place articulated Makey action figure, designed by LeFabShop.

C 3D-printed Mega Makey, designed by Daniel Spangler with Jason Babler.

D

3D DESIGN

The most challenging part of this project was designing a 3D model to make a functional robot that matched the logo from Maker Faire. To begin, I took as a reference the 2D logo originally created by Kim Dow for Maker Faire (Figure A), the 3D-printable Maker Faire Robot Action Figure (thingiverse.com/thing:331035) by LeFabShop (Figure B), the pepakura version by Rob Ives (makezine.com/projects/posable-papercraft-makey-mascot), and whatever other references I could find online (Figure C) to make sure my design was as close as possible to the character.

I started 3D modeling Makey in March of this year. I started from zero completely — from a basic cube for the body. After multiple revisions and one messed-up looking robot, I settled on its current version (Figure D).

Not only was I careful to keep Makey's overall aesthetic, I also tried to make it move in a way consistent with its natural shape and possible articulations. I had to guess how its arms and feet should move, which was a challenge. I also made the head rotate, because it seems obvious that a semi-sphere shape would rotate, and added lights to the eyes. I couldn't make the head rotate fully 360° or the wires for the eyes would break, so I decided a 180° servo would be fine for the first try.

HOW IT WORKS

Makey walks, dances, and makes sounds using the same brain as Otto — the same Arduino libraries, code (with small modifications), and microcontroller. It's also got room for small sound and touch sensors and a Bluetooth module, like the Otto DIY Plus version.

Makey uses four servos for legs and feet, same as Otto, but with the addition of two 360° servos for the arms and a 180° servo for the head (Figure E).

Finally, Makey has two LEDs for eyes to make it look like it's alive; these also serve as feedback to show that the robot is powered on.

BUILD A REAL MAKEY ROBOT

This build is new and evolving, with your help! Join us at wikifactory.com/+OttoDIY/makey for updates.

1. FABRICATION

Download the 3D files at wikifactory.com/+OttoDIY/

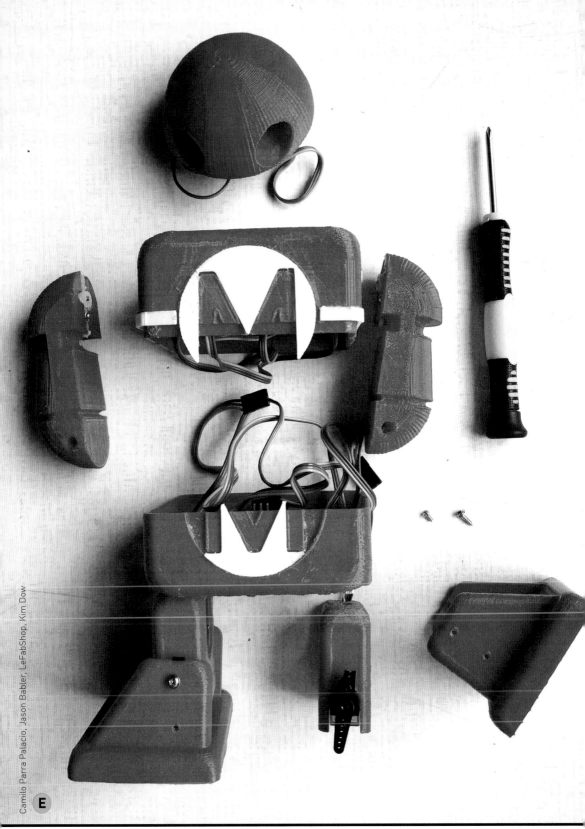

E

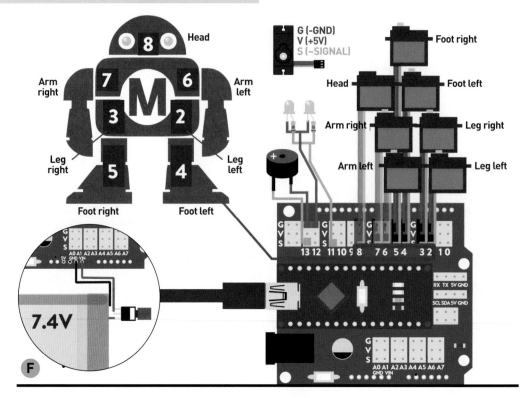

makey/files. In red, print two arms, two legs, two feet, the head, body bottom, and body top. In white, print one M logo and band for Makey's chest.

I recommend you use an FDM 3D printer with PLA material for all parts. Print them at 0.2mm resolution or less, and use 20% infill for all but the head and arms; those can be less than 10% infill, since they don't need to be strong. I use a raft to help the parts stick to the bed, but this is optional.

2. ELECTRONICS AND PROGRAMMING

It's good practice to test your electronics and software before assembling the whole robot, to avoid having to disassemble it to fix something.

Connect Makey's servos, buzzer, LEDs, and resistors to the Nano shield as shown in Figure **F**.

On your computer, download and install the Arduino IDE software. Version 1.8.5 has been the most stable. Download the Arduino libraries and sample sketches from wikifactory.com/+OttoDIY/makey/files/Firmware, then copy the libraries to *C:\Users\user\Documents\Arduino\libraries* (or wherever your libraries folder is installed).

Connect your Arduino Nano to your computer via USB, then open the Arduino IDE. Your computer

should install the drivers, and you should find your board connected in the Tools→Port menu. (If your computer didn't recognize the USB device, install the driver CH340 for your operating system.)

Now plug the Arduino into the Nano shield.

Finally, open one of the sample sketches, such as *Makey_dance.ino*, and upload it to your Arduino. If all's good, all your servos should be moving. You're ready to build your Makey.

3. ASSEMBLY

Start with the bottom parts. The foot servos connect the feet to the legs. The boards go into the body bottom with the two leg servos on either side (Figure **G**). Then you connect the legs to the body. For detailed assembly diagrams, see wikifactory. com/+OttoDIY/makey.

Prepare the body top by screwing a servo horn on top (for the head). Then put the arm servos into the body top (Figure **H**), followed by the batteries.

Prepare the head by mounting the LEDs and a 180° servo inside (Figure **I**). Thread the LED wires into the body top, then connect all the electronics as shown in Figure **F**. Connect the 7.4V LiPo and switch to Vin on the Nano shield.

G

Optionally, add sensors and Bluetooth now too.
Finally, attach the arms and head. That's it! Makey is ready to shake it.

MAKEY LIVES!

Makey's moves include classic Otto emotions and steps like the moonwalk, plus new arm and head motions and flashing eyes (Figure **J**). When we exhibited Makey at Prague Maker Faire 2019 we won a Maker of Merit award (Figure **K**)!

You can program Makey easily using drag-and-drop code blocks in Scratch, or more directly in Arduino. And if you added Bluetooth you can choreograph Makey via a free smartphone app.

REMIX IT

Makey is just one remix of Otto; there are lots of mutations, homages to famous characters, even a BoB-Otto hybrid by Kevin "The Bobfather" Biagini! Submit your version of Otto or Makey to our Otto Remix Challenge through November 3 at ottodiy.com/blog/challenge, or share it anytime at wikifactory.com/+OttoDIY. All Otto remixes retain our Creative Commons CC-BY-SA license, so the world will have free access to your remix. ◐

H

I

J

K

Camilo Parra Palacio

Join the Otto Builder Community:
ottodiy.com/#join-us

DIY Robocars:
So Ready to Rumble

Build an affordable, autonomous racer that can (finally?) beat human drivers

Written by Chris Anderson and Adam Conway

A couple years ago a few of us who had pioneered the DIY Drones movement a decade earlier decided it was time to do the same thing with self-driving cars.

Like drones before them, self-driving cars were cool but out of reach — legally, financially, safely, technically — for most people. Cars are also technically harder than drones because they need to steer precisely in a potentially crowded ground world with computer vision, unlike drones, which can navigate mostly empty sky with simple GPS. But the time was right: An explosion of new sensors, processors, and software technologies (Figure Ⓐ) had made it possible for regular people to participate in the self-driving car revolution for less than $250 and a weekend's work to set up a foot-long car that works just like the full-size ones but can be safely run indoors.

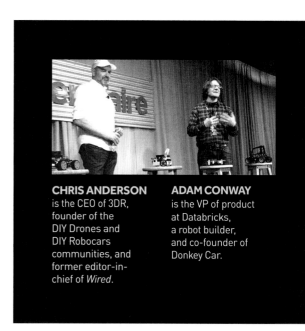

CHRIS ANDERSON is the CEO of 3DR, founder of the DIY Drones and DIY Robocars communities, and former editor-in-chief of *Wired*.

ADAM CONWAY is the VP of product at Databricks, a robot builder, and co-founder of Donkey Car.

Ⓐ BREAKTHROUGH TECHNOLOGIES

HARDWARE	EXAMPLE	ENABLES...
Sub-$200 visual odometry (VO) and SLAM (simultaneous localization and mapping)	Intel RealSense T265 tracking camera, D435 depth camera	Localization, navigation
Sub-$100 Lidar range scanning	Slamtec RPLIDAR A1	Obstacle avoidance
Sub-$100 CUDA core CPU	Nvidia Jetson Nano	Better deep learning
Sub-$50 Linux single-board computers (SBCs)	Raspberry Pi 3+	Linux for all!
Sub-$50 integrated computer vision (CV) camera	OpenMV, JeVois, Pixy2	Super-simple computer vision
SOFTWARE		
Deep learning AI	TensorFlow/Keras	Machine learning
Photorealistic simulation	Unity	Much faster development
Modern robotics frameworks	Nvidia Isaac	Much faster development, sensor fusion

Chris Anderson, Hep Svadja

The "DIY" Rubicon had been crossed once again.

Since then, the DIY Robocars movement has taken off and the community now numbers more than 10,000 people in 60-plus groups around the world. Companies such as Amazon and Nvidia have released their own AI-driven robocar kits, and robocar races are now a fixture at the leading AI developer conferences such as Google I/O and Arm DevCon.

We're getting better and faster, too. Figure **B** shows the results from our regular Bay Area race, which is now held quarterly at Circuit Launch in Oakland, California, and regularly attracts more than 100 competitors and 300 spectators (the free Brazilian BBQ doesn't hurt!). As you can see, both the computer vision and deep learning approaches are approaching the speed of our fastest human drivers — the latest winning time, posted by Mark Liu's customized Donkey Car in June, was within 0.4 seconds of the best human time!

We'll begin head-to-head human-AI races in September, and within the next six months, we predict, our homegrown robocars will be faster than humans and able to reliably race at a scale speed of 150mph (if that sounds scary, remember these are just 1/10 scale cars running at 15mph).

GET STARTED!

For high school students and below, the Minimum Viable Racer (diyrobocars.com/a-minimum-viable) is the best way to start — it's a $90 project using the fantastic OpenMV computer vision camera module that works right out of the box but can expand to deep learning. You can compare it to some other robocars in the chart in Figure **C**.

For those who want to dive right into AI, Donkey Car (donkeycar.com) is the recommended DIY kit, but if you want to buy something pre-built, Amazon's DeepRacer (amazon.com/dp/B07JMHRKQG) comes ready to run. And over the next few months, several other ready-to-run robocars are coming out, including Zümi (robolink.com) and Nvidia's JetRacer (github.com/NVIDIA-AI-IOT/jetracer).

Even if you're not ready to build or buy your own robocar, you can get involved:

- Attend our San Francisco Bay Area race/hack days: meetup.com/diyrobocars
- Join one of the dozens of other Meetup groups near you: diyrobocars.com/local-meetup-groups
- Join us on the Donkey Slack channel: donkey-slackin.herokuapp.com

B DIY ROBOCARS RACE RESULTS

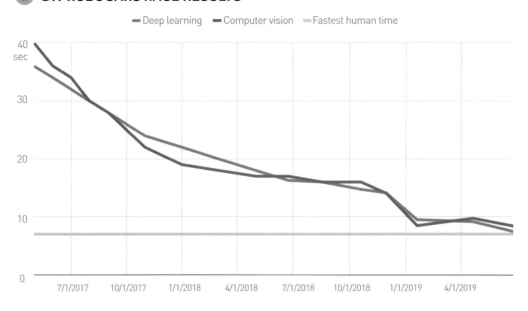

— Deep learning   — Computer vision   — Fastest human time

Ⓒ KNOW YOUR ROBOCARS

	COMPUTER VISION		DEEP LEARNING		
Type	OpenMV	Custom	Behavioral cloning	Reinforcement learning	
Standard platform	**Minimum Viable Racer**	**ConeSLAM** github.com/ a1k0n/cycloid	**Donkey Car**	**Amazon DeepRacer**, Nvidia JetRacer	
Hardware	OpenMV	Nvidia Jetson Nano	Raspberry Pi	Intel x86, Nvidia Jetson Nano	
Cost	$90	$200–$400	$250	$300–$400	

WHY RACE DIY ROBOCARS?

For a century, the car industry has innovated through racing; most of today's car tech got its start on the tracks of Formula 1 or Monte Carlo. But with autonomous cars, it's been mostly the opposite: driving slowly and cautiously. That's why the semi-autonomous cars on the streets today drive like little old ladies. Our hope is that our sub-scale, no-passenger approach to autonomy may reveal a different path to performance and safety, one more about nimbleness and aggressive avoidance of danger.

In short, just as the Homebrew Robotics Club created Apple, which started making the worst computer you could buy and ended up making the best, we're betting a homebrew approach to self-driving cars could inspire the same. ◉

Build your own autonomous racer with our Donkey V2 tutorial at makezine.com/projects/build-autonomous-rc-car-raspberry-pi

Soap Vomiting Unicorn!

Written and photographed by Britt Michelsen

Hack a cheap automatic soap dispenser to build your own wash-time buddy

If you ever thought cleaning would be infinitely more fun if it could be done with unicorn vomit, you've come to the right place. I'm going to show how to 3D print and build your own soap dispenser that looks like a unicorn.

The horn can be unscrewed to add soap (Figures **A** and **B**). There's a pushbutton hidden in the right nostril that turns the sensor on and can be used to increase the amount of soap dispensed. Another pushbutton hidden in the left nostril turns the sensor off or reduces the amount of soap dispensed. The battery compartment is hidden in the right foot.

I'd wanted to make something with a unicorn for quite some time. Originally I planned to build a wine bottle opener but decided against it, since I don't drink wine. I always loved Sam Elder's Pooping Reindeer Candy Dispenser on Instructables, and thought about turning it into a unicorn, but decided that since his version is already so great, turning it into a unicorn wouldn't add much (and I'm allergic to chocolate). Finally, after walking into my kitchen and seeing my Deadpool Knife Block, I decided that a unicorn vomiting Ajax would go perfectly with it.

1. 3D DESIGN

I tried to find a unicorn I could use but sadly, the one I really liked was way too small. So I decided to design and 3D print my own version.

I wanted the electronics to be hidden as well as possible, so I went through a few design iterations. The unicorn was designed in Autodesk Fusion 360. It was my first time trying the Sculpt feature and I absolutely loved it. I used it for the head, horn, ears, and mane. Simply select Create→Create Form and you'll be in the sculpting environment, where you're able to create forms and edit points

TIME REQUIRED:
A Weekend

DIFFICULTY:
Intermediate

COST:
$40–$80

MATERIALS
- » **"Touchless" liquid hand soap dispenser with motion sensor** such as Amazon #B0789KF36W or B077QBF1WP. These are widely available under different brand names; I used the type with an LED in the top, not at the bottom.
- » **Battery holder, 4xAAA** Amazon #B00QLQQI58
- » **3D printed parts** Download the free STL files at instructables.com/id/Soap-Vomiting-Unicorn and print them out, or send them to a service for printing.
- » **Magnets, round, 6mm dia. × 1.5mm thick (12)**
- » **Hookup wire**
- » **Heat-shrink tubing**
- » **Zip tie**
- » **Silicone tubing, 5mm OD, 3mm ID, about a 30cm length**
- » **Two-part epoxy**
- » **Hot glue**
- » **Auto body filler putty**
- » **Primer**
- » **Paints: white, gold, and your choice for the mane and tail**

TOOLS
- » **3D printer**
- » **Soldering iron**
- » **Screwdriver**

A

B

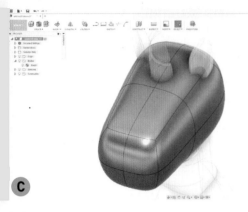

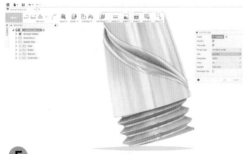

as well as edges (Figure **C**). This way you can design more complex objects than in the Model feature. I highly suggest you give it a shot.

Once I was happy with the shape of the unicorn (Figure **D**), I cut off its snout (wow, that sounds harsh) and hollowed the unicorn by using the Shell tool. I left out the snout because I didn't want to turn it into a shell; I later cut slots into it to fit my dispenser's pump and pushbuttons. If you use a different dispenser, you may need to adjust the design to fit.

> **TIP:** If you're making something hollow, do so *before* using fillets. I had a few failed attempts with the hollowing feature not working properly. Once I figured out it was because of the fillets, it worked perfectly.

Next I designed a divider so the soap wouldn't get in contact with the electronics (Figure **E**), and I placed the battery compartment in one of the feet. As you might have guessed, the threads were designed with the Thread feature, which works really well (Figure **F**). All that was left to do was to change the inside so that the electronics would fit into it.

I decided to hold the feet and the snout in place with magnets. I was a bit nervous that water might get into the battery compartment, but so far it has worked really well.

2. 3D-PRINT THE PARTS

To print the body (Figure **G**), I used a Support Overhang Angle of 70°. This way there aren't any support structures in the middle of the soap dispenser. The ones in the top thread can easily be removed.

Print each part once, except *Nostril.stl*; print that one twice. I used a layer height of 0.15mm in order to save time (Figure **H**), and I knew I would smooth the parts later anyhow.

I printed everything with a 0.4mm nozzle. Depending on your printer, you might have to print *TubeConnector.stl* using a finer nozzle to get a good fit with the silicone tubing.

3. SMOOTH THE PARTS

To smooth the unicorn, I started by filling the bigger uneven areas with auto body filler (Figure

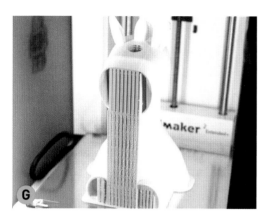

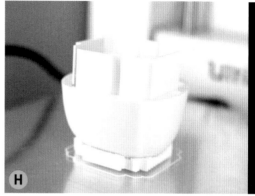

(I). Once it was cured, I wet-sanded the parts I had filled. After cleaning them I applied two more thin layers of auto body filler and wet-sanded everything again (Figures J and K). To learn more, check out my tutorial at instructables.com/id/How-to-Smooth-PLA-3D-Prints.

I also taped off the thread of the horn to make sure that it would still fit later on.

After you've smoothed all the parts, glue the magnets in place.

4. PAINT THE BODY

I used a white primer on everything except the horn, and then airbrushed the parts with Createx Wicked Colors White (Figure L).

Then I ran into troubles. Since I was too lazy to tape everything, I decided to use a brush to paint the mane, but the paints I chose didn't work too well (Figure M).

So I decided to paint it white again and tried using masking putty (Figure N on the following page), but that didn't work too well either. I couldn't

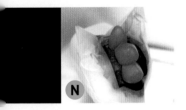

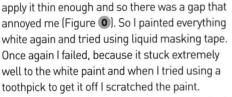

apply it thin enough and so there was a gap that annoyed me (Figure **O**). So I painted everything white again and tried using liquid masking tape. Once again I failed, because it stuck extremely well to the white paint and when I tried using a toothpick to get it off I scratched the paint.

Finally, I decided to paint the mane and tail with a high-pigmented pink paint, which I could apply with a brush so I wouldn't have to mask anything!

5. PAINT THE HORN AND EYES

Before painting the horn I covered the thread with tape to make sure it would fit later. In my experience, metallic paint ends up looking a lot better if you prime the parts with black paint, so I used The Army Painter Matte Black primer (Figure **P**) and then Createx Pearlized Satin Gold airbrush paint, which I love (Figure **Q**). After it dried I used a glossy varnish to protect it.

It was hard for me to decide how big I wanted the eyes to be and where to position them, so I printed a few in different sizes and held them to the side of the unicorn (Figure **R**). I ended up going with 8mm (0.315") eyes placed 5.5cm apart. As you can see I went a bit overboard and used my plotter to cut a stencil (Figure **S**).

6. DISASSEMBLE THE DISPENSER

Disassembling the soap dispenser is quite easy. Just remove the screws that are hidden in the battery compartment and the screw that's next to the opening where the soap is dispensed.

Then remove the two screws holding the PCB (Figure **T**) and the screws holding the motor. Once you've done that you can pull all the electronics out of the metal tube (Figure **U**).

7. ASSEMBLE THE SNOUT

Unsolder the LED next to the pushbuttons. Then make sure the two 3D-printed *Nostril.stl* pieces fit onto the pushbuttons and into the nostrils. You'll have to turn them to the outside, as in Figure **V**.

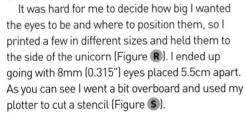

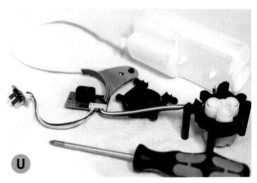

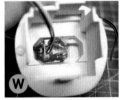

Place the prints and the PCB into the snout and secure the PCB in place with hot glue (Figures **W** and **X**). Be careful and don't use too much hot glue, since heat and 3D prints tend to not mix well.

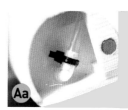

8. ASSEMBLE THE FEET

Extend the power wires if necessary by soldering short lengths of hookup wire. Push the wires through the head, pull them through the foot, and then solder them to the battery compartment (Figure **Y**). Now glue the battery compartment to the left foot, *Foot02.stl* (Figure **Z**).

Next, use *TubeConnector.stl* to connect the hole in the right foot to the silicone tube, as shown in Figure **Aa**. Zip-tie the tube tightly onto the connector, then glue the connector into its hole.

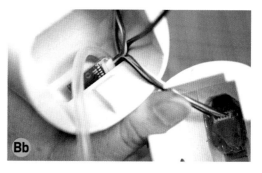

> **CAUTION:** Don't use hot glue here, it will melt the tube connector — believe me, I tried! Use epoxy.

9. CONNECT THE PUMP

Push the silicone tube through the body and the sensor PCB into its hole (Figure **Bb**). Admittedly this is a bit fiddly.

Then solder the pump motor wires back in place (Figure **Cc**). Make sure you get the polarity right.

Place the silicone tube into the pump body (Figure **Dd**), screw it closed, and push the end of the tube though the unicorn's mouth.

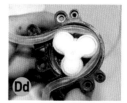

Now put the pump into the snout (Figure **Ee**) and pull the silicone tube further through the mouth (Figure **Ff**). Test your unicorn, then trim the tube.

Congratulations, you're done!

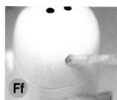

PUKE AND SCRUB!

Positioning your unicorn is key, otherwise it will start dispensing soap even though you just wanted to get some water or somebody walked by too close. If I were to redesign it I'd make the assembly a bit less fiddly, but otherwise I'm happy with it. Even after a year it's still fun to use and a great conversation starter. ⊘

BRITT MICHELSEN is a chemical engineer in Hamburg, Germany, interested in computational fluid dynamics. To balance all the theoretical work, she likes to make stuff in her free time.

See more fun projects from Britt Michelsen at instructables.com/member/BrittLiv.

DIY Inflatable Car

How I built an amphibious vehicle from hacked power tools

Written by Axel Borg

I'm always looking for an excuse to play outdoors as an adult, especially where land meets water. I love being out in nature, and love creating tech stuff, so here I am combining those favorites in an ultimate everyday-object hack — an amphibious, DIY inflatable car made from power tools and vacuum cleaners.

I was inspired by the amazing Russian-made Sherp amphibious ATV, but mine is nothing like that. I'm not sure how to describe it — it's like a cross between a bouncy castle and an ATV, with a fun factor to match! The whole vehicle only weighs 130kg (287lbs) set up and ready to go (Figure A).

DIY INFLATABLES SECRET SAUCE

I've been looking for a DIY way of creating inflatable airtight structures for quite some time and got a break when I stumbled onto an extremely adhesive tape typically used in house construction to permanently seal and bind plastic foil vapor barriers together. I used Etab #5525; 3M makes an equivalent, #8067E FAST-F. This way you can sew together your structure with a ordinary sewing machine (Figure B) and then create an airtight

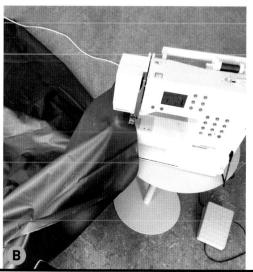

A

B

Axel and Sara Borg

TIME REQUIRED:
Many Weekends

DIFFICULTY:
Advanced

COST:
$1,000–$2,000

MATERIALS
» **Heavy duty tarpaulin, laminated PVC, 600g/m² weight** for sewing the inflatable wheels
» **Professional vapor barrier sealing tape** to seal the seams, such as Tesa Seal Flex #60073, Etab #5525, or 3M #8067E FAST-F.
» **Wet/dry shop vacuum cleaners (2)** used in reverse as blowers for inflation
» **Power tool motors, series wound (4)** one for each wheel. DC- or AC-compatible will work. I used motors from AC immersion concrete mixers, HiKoki Niko #BM1600, because they're already geared down, but I recommend a post-hole auger for even lower gearing.
» **Cordless drill** I hacked this just to get the PWM (pulse-width modulation) motor control circuitry to drive my bigger motors.
» **Battery pack, high voltage** I connect 15 LiPo batteries in series, 14.8V each for about 230V total, equivalent to European mains power; in the USA you could use half that.
» **Transistors, high voltage MOSFET/HEXFET**
» **Resistors, 30Ω–100Ω**
» **Diodes, flyback**
» **Plywood**
» **Garden hose** for inflation plumbing
» **3D printed connectors** for inflation plumbing and vehicle frame
» **Fiberglass and epoxy** to strengthen frame connectors
» **Aluminum tubing** for frame
» **Steel chain, ½", and sprockets** for the final 1:3 gear reduction; I got them from technobotsonline.com in the UK.
» **Various parts** for braking and steering

TOOLS
» Sewing machine
» 3D printer
» Soldering iron and solder
» Electric hacksaw
» Cordless drill/driver
» Various files, cutters, pliers, wrenches, etc.

AXEL BORG is a designer in Sweden with a passion for anything mechanical. He teaches technology to kids and elementary school teachers, and shares his creative adventures on his YouTube channel, amazingdiyprojects. This project won grand prize in our Mission to Make contest for *Make:* Volume 70.

Axel and Sara Borg

seal by taping all your seams from the inside (Figure **C**).

In this build I use 600g/m^2 vinyl tarpaulin for the wheels — I did try lighter materials first, but they deformed or got shredded under early load tests in my backyard.

CORDLESS DRILL CONTROL HACK

The other DIY breakthrough for me was when I discovered how easy it was to drive high-voltage-rated MOSFET/HEXFET transistors using the PWM hardware from cheap cordless drills (Figure **D**). All you need is a 30Ω–100Ω pulldown resistor between the high-voltage transistor's gate and source leads, and a freewheel (flyback) diode across the motor's terminals to take care of the inductive spikes from the motor (Figure **E**). Given a sufficiently high-voltage battery pack, you're now ready to use pretty much any series-wound power tool you can find, in new creative ways.

SHOP VAC INFLATORS

I chose two cheap wet/dry shop vacuum cleaners (Figure **F**) for inflating the wheels because they have a separate cooling fan for their motors, rather than using the airflow from the suction. Since the wheels are pretty much airtight, the airflow will stall once they're fully inflated (Figure **G**), and the motors would overheat without the separate cooling system. Of course, you want to use a vacuum cleaner that has the option to fit the hose on the exhaust side as well, making this odd setup, using them as blowers, more convenient.

3D PRINTED CONNECTORS

Like many of my builds, this one also depends heavily on 3D-printed parts. The inflation system plumbing uses pipes designed in TinkerCAD and printed in PLA, matching the ½" garden hose that I use to route the air from the vacuum cleaners to each wheel.

For the vehicle frame, I also printed connectors to hold the 50mm aluminum tubing together (Figure **H**). For structural strength, these parts need to be reinforced with fiberglass and epoxy (Figure **I**). The 3D printer is good at producing lightweight parts with correct dimensions, and the fiberglass contributes the strength needed.

G

HIGH VOLTAGE BATTERY PACK

Since I live in Europe where we have 230V mains voltage, that's the battery pack voltage I need. You have to be careful and think a step ahead when working on a system at these voltages. I use 3D-printed main connectors carrying 15 XT60 hobby battery plugs which connect all 15 of my LiPo batteries in series *only* when the main connector is screwed together (Figure **J**). Only then is 230V present in the vehicle; at all other times the highest voltage present is 14.8V from a single 4S LiPo pack.

GEARING: HOW LOW CAN YOU GO

When using a power tool motor in a project like this, look for the most geared-down one you can find. Ideally I'd use an electric earth auger, which is geared down around 120:1. I used some concrete stirrer machines — like industrial-strength immersion blenders — that were geared down to 80:1, which isn't enough, so I added 3:1 final chain sprocket reduction to turn the 2m (6') diameter wheels forcefully enough (Figure **K** on the following page). If I build another version I'll go for the auger motor/gearbox combo. Right now, my motors run too hot when I drive slowly; ideally I'd have a cooling system that runs at full blast even when the motors are not.

H

I

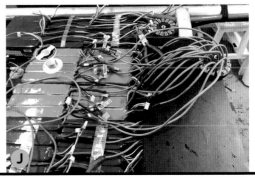

J

STREET

TURF

SAND

WATER

Axel and Sara Borg

FLYING OVER WATER

Driving the vehicle out onto the water for the first time was unreal — just absolutely wonderful! The feeling is comparable with taking off in an airplane, just so much slower. The top speed on water is very low, just a couple of km/h, but the traction is great, over 100kg of propulsive force! On land the top speed is well over 30km/h — more than practical, given the limited vision while sitting down and driving it.

I'm looking forward to driving my kids to school in this friendly beast so that they can maintain their open mindset for life! ●

AHA! MOMENT

A big grin showed up on my face the first time I spun up a ordinary corded electric drill to full power from a 240V LiPo battery pack via a cordless drill I just hacked! Those small series-wound motors that we use in almost every plug-in household tool (drills, vacuum cleaners, food processors, etc.) really rock!

UH-OH! MOMENT

There were vertical oscillations when steering the vehicle like a tank — totally unexpected. Pivot steering fixed it.

Also, some time-consuming problem solving related to inductance issues when hooking up several motors to one battery pack = sharing common ground. In the end I made a workaround: I divided the battery packs into three subassemblies; one for each side providing electricity to two in-wheel motors, and one for the two inflating vacuum cleaner motors. This reduced the noise enough to keep the transistors happy and to keep the central cordless drill PWM hardware hack simple.

Check out this project and share your ideas at makershare.com/projects/inflatable-car-sort, and watch the whole video series at youtube.com/user/amazingdiyprojects.

Make:
Community

CONNECTING THE PEOPLE AND PROJECTS THAT SHAPE OUR FUTURE

Make: Community offers a place to **connect**, **collaborate**, and **form our community**. It is a melting pot where makers, makerspaces, content creators, teachers, students, businesses, and those just dabbling in the world of ingenuity and learning come together to support the Maker Movement, which is a driving force of our future.

As a member you can access and interact with thousands of like-minded makers from around the world — we have members from over 50 countries already!

Make: Community membership includes:
- Digital magazine subscription
- Discounted print subscription
- Directory of makers, members, and makerspaces
- Community platform and groups
- Exclusive videos
- Online video chat forums with staff and guest MC's
- A voice in the direction and causes of Make: Community

Chris Willis

In Your Face!

Written and photographed by Forrest M. Mims III

FORREST M. MIMS III, an amateur scientist and Rolex Award winner, was named by *Discover* magazine as one of the "50 Best Brains in Science." His books have sold more than 7 million copies. He's based in Texas. forrestmims.org

Measure your face's exposure to UV rays with "Sunny Sam"

Too much ultraviolet radiation can cause sunburn and, over time, skin cancer. The EPA's UV Index (UVI) provides an important way to keep track of your sunshine exposure.

The UVI works well for horizontal surfaces, such as your back and the backs of your legs when you're lying facedown on a beach. But our bodies are complex in shape, and the UVI may not apply for much of your body when you are walking along the same beach.

Last summer I spent 28 days on Hawai'i Island studying the intensity of UV at the eyes, ears, and cheeks of a simulated human face named "Sunny Sam." The results were surprising, for when Sunny Sam was looking forward when the sun was straight overhead at noon, his face received much less UV than at midmorning or mid-afternoon. This suggests that the UV Index needs clarification so people will better understand the risks they face when outdoors on sunny days.

BUILDING A SIMPLE SUNNY SAM

My original plan to measure solar UV on a human face was to wear a Phantom of the Opera-style half mask equipped with several UV sensors. I tried this approach and it worked (Figure Ⓐ). It allows you to measure UV under a wide range of realistic conditions. But the mask terrified children and even some family members. When I posted a photo of me wearing the mask on Facebook, dozens of comments quickly appeared, including these:

- *Forrest, this borders on being bizarre!!*
- *I feel that painting the mask flesh-toned could reduce the freak-out factor of observers.*
- *"We are the BORG. Your DATA will be assimilated. Resistance is futile."*
- *The Phantom of the Forrest.*
- *Part man, part machine*
- *You look like one of the Borg on* Star Trek.

TIME REQUIRED:
2–5 Hours

DIFFICULTY:
Easy to Intermediate

COST:
$25–$300

MATERIALS
BASIC VERSION:
- » **Mannequin head** from a craft store
- » **Analog UV Sensor Board with GUVA512SD sensor (1 or more)** Adafruit #1918, adafruit.com
- » **Wrapping wire** such as 28 or 30 AWG wire gauge
- » **Battery, 3V** either a lithium cell or 2 AAA cells in a holder
- » **Voltmeter**

ADVANCED VERSION:
- » **Data logger** Onset 4-channel, 16-bit logger, onsetcomp.com
- » **UVB photodiode, 220nm–320nm** Roithner #GUVB-T21GH, roithner-laser.com/pd_uv.html
- » **PVC pipe scrap, ¹⁷/₃₂" OD, ¹¹/₁₆" long**
- » **Teflon disc, ½" diameter, about 0.5mm–1mm thick** I used Cox #49DDISC, coxengines.ca. You can also use Teflon film from a sewing supply store.
- » **Turntable, battery powered** I used Master Tools' 7-inch TurnTable.
- » **Pedestal** to mount head to turntable, from a craft store
- » **Tripod**
- » **Bubble level**

TOOLS
- » **Wire wrap tool**
- » **Wire cutters**
- » **Soldering iron and solder** preferably a low-power USB iron

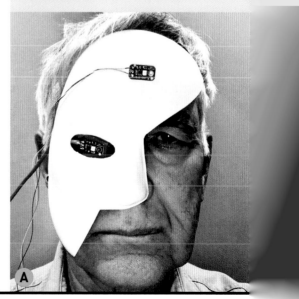

Ⓐ

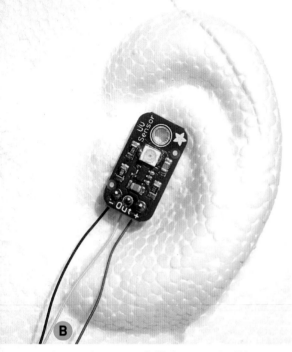

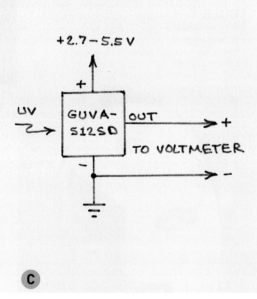

B

Inexpensive sensor suitable for measuring UV at the face and other parts of the anatomy.

+2.7 – 5.5 V

UV → GUVA-512SD OUT → + TO VOLTMETER → –

C

Basic circuit for operating the UV sensor in Figure B.

The face mask approach has potential for applications in which you measure the UV at your face while you're looking in various directions. But I decided another approach might be best for public measurements: I used a foam plastic mannequin head for mounting the UV sensors where they will be partially shielded by facial anatomy.

A low-budget Sunny Sam can be made by using adhesive tape or velcro to attach inexpensive UV sensor breakout boards directly to various places on the head. Figure **B**, for example, shows an Adafruit #1918 UV sensor installed at Sunny Sam's left ear. These UV sensors are described in detail in my previous article "DIY Sunburn Sensors" (*Make: Volume 65, pp. 62–64*, makezine.com/projects/build-hack-and-deploy-detectors-to-measure-solar-uv-radiation). A 3-volt battery (lithium or two AAA cells) to power the sensor(s) can be taped to the head's neck. Connections between the battery and the UV sensor (Figure **C**) can be made with wrapping wire or other small gauge wire.

The output of each sensor can be indicated by a digital voltmeter. You can then measure the UV exposure at the sensor by manually adjusting the orientation of the head with respect to the sun. Describe your readings into a tape recorder or smartphone and transcribe them later.

A FULLY AUTOMATED SUNNY SAM
The Sunny Sam used for the Hawai'i UV survey was equipped with modular UV sensors (GenUV GUVB-T21GH from Roithner) having a closer spectral response to the *erythemal* (sunburning) wavelengths of UV than the breakout board described above. The *Make:* article cited above describes how to mount these sensors in a holder fitted with a Teflon diffuser cap.

I made seven of these sensors for Sunny Sam and equipped each with three 8" wrapping wire leads soldered to ⅛" phone plugs that can be inserted into sockets on an Onset 16-bit data logger. The leads were folded against the sides of the sensors, and the sensors were then inserted into NIBCO 4701 Series CPVC pipe couplers (1⅛"×⅞") placed in holes formed at the eyes, ears, cheeks, and forehead of the mannequin head. To make the holes, I pressed a thin-wall, 1"-diameter aluminum tube from a turkey baster 1½" into the head, rotated the tube, and removed it together

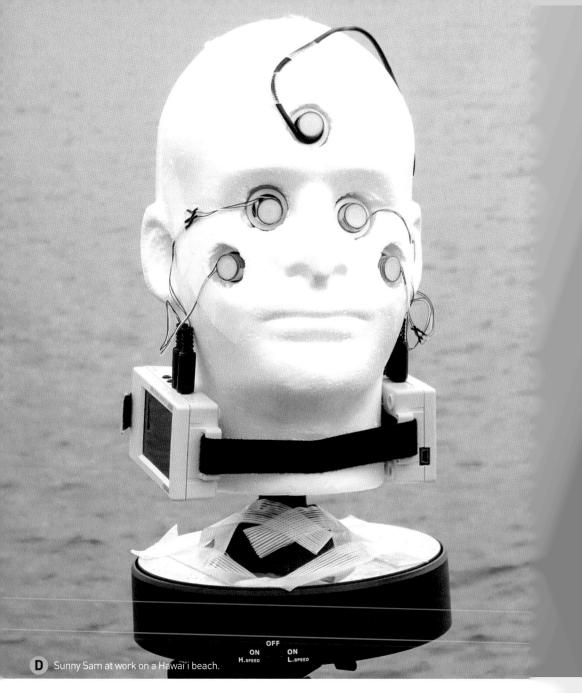

OFF

ON
H.SPEED

ON
L.SPEED

D Sunny Sam at work on a Hawaiʻi beach.

with a cylinder of foam plastic.

A pair of Onset 4-channel, 16-bit loggers was attached to the neck using velcro. The sensor phone plugs were then inserted into the logger's jacks. The loggers were then programmed to switch on at the desired time and record data once per second.

A battery-powered turntable used to display jewelry was attached to the top of a tripod by inserting the tripod's camera screw through a hole bored through the turntable's battery cover. A bubble level was placed atop the turntable, which was adjusted to be perfectly level. The mannequin head was then placed over a plastic pedestal attached atop the turntable. When the turntable is switched on, the mannequin head rotates once every 36 seconds or 100 times per hour. Sunny Sam in operation is shown in Figure **D**.

The author (left) and Sunny Sam discuss the very high UV at Hawai'i's high-altitude Mauna Loa Observatory.

RESULTS

During July and August of 2018, Sunny Sam was installed at seven locations on Hawai'i Island in a 28-day study sponsored by Rolex. Most sites were at or near sea level, and one was at the high-altitude (11,200 feet) Mauna Loa Observatory (Figure **E**). No site was perfect, but five had a good view of most or nearly all the sky.

Why Hawai'i? At about 19°N latitude, the island is below the Tropic of Cancer (23.4°N). This means that on two days of the year, the sun is directly overhead at noon, and upright objects like vertical poles (and people) cast no shadow. Noon in Hawai'i on these special days is called *Lāhainā noon*.

Sunny Sam sessions on or near the July 24 Lāhainā noon provided the widest possible range of sun elevations: 0 to 90 degrees over the horizon.

Tests in Texas a week prior to Hawai'i showed that Sunny Sam's face received much less exposure to UV at solar noon than at mid-morning and mid-afternoon. This is because the face sensors are pointed sideways and not upward. This is not predicted by the UV Index, which is based on the UV falling on a horizontal surface and which always gives noon as the highest UV exposure.

The Hawai'i results were even more dramatic, especially during the noon hour on days near Lāhainā noon. Figure **F** shows results from July 22 at the Old Kona Airport Park beach. The maximum sideways UVI at the left ear of a simulated person walking along the beach at noon was only 3.5, while the horizontal UVI on the

UV Index at Old Kona Airport Park Beach (July 22, 2018)

UV Index at Old Kona Airport Park Beach (July 22, 2018)

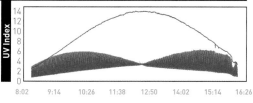

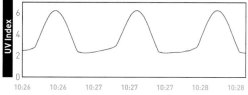

HAWAII STANDARD TIME

— UV Index (zenith)   — UV Index at left ear

HAWAII STANDARD TIME

— UV Index (zenith)   — UV Index at left ear

 F Full day results for horizontal UV Index and UVI at the left ear of Sunny Sam.

 G Three rotations from Figure D during peak UV at mid-morning.

H 190-degree image of what a side-facing UV sensor views at the Mauna Loa Observatory.

back of a sunbather lying flat on the beach was an extremely high 14.

Figure **G** shows the UVI at Sunny Sam's left ear during three rotations of the head at mid-morning when the peak UV at the ear was 6 and the UV Index was 10.

GOING FURTHER

Sunny Sam would make a great science fair project. (Be sure to acknowledge this article.) For example, the steep fall in UV intensity at solar noon measured in Hawai'i will be reduced at higher latitudes, since the sun is lower in the sky at noon and the side-viewing sensors receive more direct sunlight.

There is very little in the scientific literature about the angular response of UV sensors pointed sideways (Figure **H**), when half the UV is from the sky and half from what's below the horizon (reflected off pavement, water, soil, plants, etc.). Perhaps you can contribute new knowledge.

For more information, see Sunny Sam's Hawai'i results in "Measuring and Visualizing Solar UV for a Wide Range of Atmospheric Conditions on Hawai'i Island," a paper by several UV experts and me published in International Journal of Environmental Science and Public Health (2019, doi.org/10.3390/ijerph16060997). Among the co-authors are Prof. Andrew McGonigle of the University of Sheffield and his Ph.D. student Thomas Wilkes, both of whom accompanied me on the Hawaii phase of this study. ⊘

Play That Funky Music

Turn a vintage radio into a lo-fi guitar amp

Written by Marcus Dunn

This is probably one of my favorite hacks that I've ever stumbled across! It's so simple that anyone with a soldering iron and screwdriver can do it.

All radios have an amplifier built into them — it's how you're able to increase the volume. This project taps into the radio's amplifier so you can play guitar through it.

Hacking a radio can give you the most amazing vintage tone. The distinct "low-fidelity" sounds that come out of these vintage radios will really surprise you. It's a raw and dirty sound perfect for playing bluesy riffs. There's also the added bonus of creating your very own tone as no two radios are ever alike.

There's a good chance that the radio you'll use for this has seen some miles. The speaker might hiss or rattle, the volume could jump around, or it has some other left-of-center feature, which is what makes these amps so cool. You never know what sound you're going to get.

I've made a few of these now and this version has added filters (in the way of capacitors) so the sound is even better. You can swap from clean to low-down-dirty at a flick of a switch.

1. FIND THE RIGHT RADIO

Some radios work better than others. Look for one with at least a 3" speaker (Figure **A**). This will ensure that your speaker is big enough to have an effect on the tone.

The size and quantity of the batteries that the radio takes affects the tone too! The bigger the batteries, the fuller the tone, and the more volume the radio will have. Find a radio that takes C- or D-cell batteries for the best results. The radio I used takes 6 D batteries, so it has plenty of power to drive the 5-amp speaker (Figure **B**). You can modify small transistor radios powered by 9V or AA batteries, but I've found they aren't very loud and don't have enough distortion for me.

If your radio has extra controls like bass, tone, or treble, then you're in radio amp heaven. These will give you extra ways to change the sound and tone of your amp.

2. OPEN UP YOUR RADIO

To do this mod, you'll need to be able to get to the radio volume potentiometer. This hack won't affect

TIME REQUIRED:
1 Hour

DIFFICULTY:
Beginner to intermediate

COST:
$5–$7 (not including radio)

MATERIALS

» **Vintage radio** If you don't have one lying around, then try a thrift store, secondhand shop, or eBay.
» **Nichicon FW audio capacitor, 220μF, 50V** This is a good quality audio cap. You can use a cheaper one if you want to but it might affect the sound quality.
» **Capacitor, polyester, 100nF, 100V**
» **Switch, SPDT**
» **Potentiometer, 10K**
» **Mono input jack, ¼"**
» **Prototype board**
» **Wire**

TOOLS

» **Soldering iron**
» **Screwdriver, Phillips head**
» **Drill**
» **Tape or hot glue gun**
» **Guitar and cord**

A

B

Marcus Dunn

MARCUS DUNN
is a maker and explorer of all things. He currently lives in Melbourne, Australia, and has a nasty habit of pulling things apart.

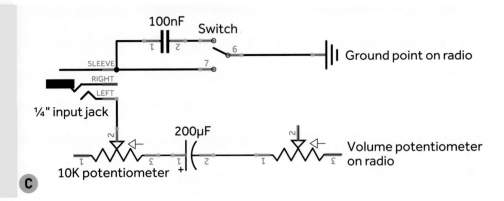

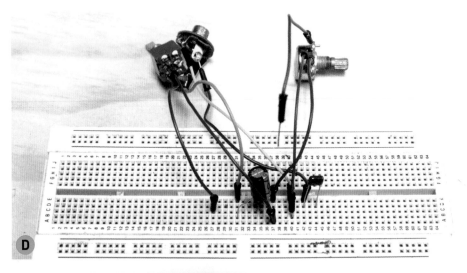

C

D

the radio's ability to play music either — you'll still be able to listen to the radio through it.

First, open the back of the radio. Flip the radio over and remove the screws holding the back into place. With a screwdriver, lever off the knobs and switches. They should pop off with some jiggling.

Gently remove the cover making sure that no wires are pulled out. You should be able to see the volume potentiometer and also the solder points.

3. BREADBOARD THE CIRCUIT

The circuit is pretty simple to make, but it's best to test it first on a breadboard following the schematic (Figures **C** and **D**). You may have to change the value of the capacitors for your radio. I've included the Fritzing file at makezine.com/go/radio-hack-guitar-amp in case you want to make any changes. If you want to create your own circuit schematics, download Fritzing at fritzing.org.

Marcus Dunn

E

F

G

4. FIND THE SOLDER POINT ON THE VOLUME POTENTIOMETER

To test the circuit, you'll need to find the correct solder point on radio's volume potentiometer. Use the AM band and tune the radio so it's not on any station, and turn down the volume until you don't hear any static. You'll need to connect a jumper wire to a ground point on the radio — I used the ground solder point on the speaker (Figure E).

With the breadboarded circuit connected to ground on the speaker and the guitar plugged into the jack, begin testing each point on the volume potentiometer with the positive jumper wire (Figure F). There are usually about five solder points on the volume potentiometer, and the wire from the jack needs to be soldered to the right one for the amp to work. Place the jumper wire against the first solder point on the volume potentiometer and strum the guitar. If you hear nothing, move on to the next one. Once you find the right one (you'll hear the guitar through the speaker), solder the jumper lead to it.

Experiment with different capacitor values if necessary until you're happy with the sound.

5. SOLDER THE CIRCUIT

Now it's time to make a more permanent circuit.

Following either my schematic or the new schematic you've developed for your particular radio, rebuild your circuit using the prototype board (Figure G). Make sure the wires for the 10K potentiometer, switch, and jack are longer than necessary — it's easier to trim than to add.

6. ADD THE COMPONENTS TO THE RADIO

Find space along the radio case to fit the 10K potentiometer, switch, and input jack. You'll also

need to be able to stick the circuit board down, so make sure there's enough room inside. The good thing is that most vintage radios have empty space inside that will accommodate the components.

Drill holes to fit the potentiometer, switch, and jack and secure them to the case (Figure **H**). Place your circuit board inside and make sure you can secure it in place later on.

7. CONNECT THE CIRCUIT BOARD

It's now time to connect your circuit to the components. Following the schematic, wire the switch to both the 100nF capacitor and ground on the speaker (Figures **I** and **J**). Wire the 10K potentiometer — it's used for volume control and helps filter the sound — and wire the input jack (Figure **K**). Finally, connect the wire from your circuit to the solder point on the radio's volume potentiometer.

Plug in your guitar and test to make sure the circuit is working. If everything is connected correctly, you should start to hear some sweet sounds coming from the speaker.

FOR THOSE ABOUT TO ROCK

As I mentioned earlier, the hack won't damage the radio at all and you can still use it just as a radio. Using it as a guitar amp is simple, but I thought I would add a few tips.

- Don't turn the radio volume up too loud. Start with it at zero and turn it up as needed.
- You can use either AM or FM. I don't find there is any real difference.
- Tune in the radio so it's not on a station.
- Plug the cord into the jack on the radio and then plug it into your guitar.
- Try adjusting the 10K potentiometer and strumming the guitar. Find the sweet spot for the volume. You can also turn up the radio volume as well. Just note that you might hear some background static if it's too loud. The radio I used can be turned up without any noise, but it becomes sensitive and the sound starts to decay.
- If your radio has bass and treble controls then play around with these as well. You can get some awesome sounds by adjusting them.
- Try flicking the switch. You'll find that you either have a very clean sound or a grungy, lo-fi sound.
- Finally, have fun playing your amp and experiment to see what other sounds you can get out of it. ⊘

For more photos and a video of the amp in action, visit makezine.com/go/radio-hack-guitar-amp.

I'VE GOT THE MUSIC IN ME
KEEP ON PLAYING...

The $5 Cracker Box Amp
makezine.com/projects/make-09/the-5-cracker-box-amp

License Plate Guitar
makezine.com/projects/license-plate-guitar

Optical Tremolo Box
makezine.com/projects/make-33/optical-tremolo-box

Build an Arduino Guitar Pedal Stompbox
makezine.com/projects/arduino-guitar-pedal

Marcus Dunn, Hep Svadja, Sean Michael Ragan, Andrew Goodman, Sam Murphy

BIG,
Printin'

Chop up your designs when you need something bigger than your bed

Written by Caleb Kraft

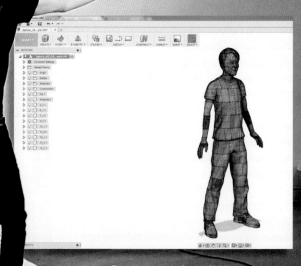

James Bruton poses in front of his 12' tall sculpture of himself, made and segmented in Fusion 360.

Cut
Z: 152.7
Keep: Upper part: ☑ Lower part: ☑
Rotate lower part upwards: ☑
Show preview:
Perform cut

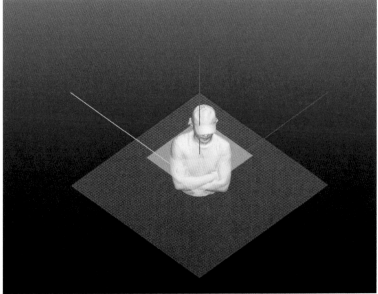

A

Sometimes you get a really big idea. An idea so big that, when printed, it will dwarf your 3D printer. This presents a surprisingly common problem — how do you print something bigger than your machine?

There are two main ways of dealing with this:

1. Design your file for printing in small chunks by including alignment features, and not exceeding the size of your printer with each part.
2. Take a file that already exists, and cut it up using various methods to get it down to the sizes you need for later assembly.

In our experience, the second method seems to be the more common approach, and is the one used by collaborative groups like We the Builders (wethebuilders.com) and The Great Duck Project (thegreatduckproject.org) to create their large scale, crowd-sourced sculptures.

Let's look at a couple options for cutting big things into easy-to-assemble smaller chunks.

Your Slicer

In many slicers for your 3D printer (like Slic3r), there are options to split the model in half on the Z axis, which can allow you a bit of flexibility. Other programs like Cura allow you to cut a file similarly by manually moving it down into the print bed. This isn't a very flexible approach, but can save you in

B

We the Rosies, designed by Jen Schachter and segmented in Netfabb and Meshmixer by Todd Blatt.

C

a pinch, especially if you have time to get creative with a manual solution (Figure **A**).

Your Engineering Software of Choice

Pretty much any CAD software is going to have the capability, with some effort required, to split parts (Figure **B** and **C**). We've seen it done with clever

Courtesy James Bruton, Jen Schachter

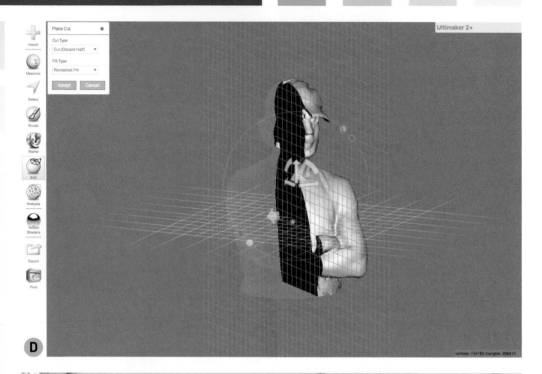

The Luban-generated *Great Duck Project* by Jesse Robinson and Nicholas Iacobelli, made for Maker Faire Westport.

scripts in OpenScad, and manually in Autodesk Fusion 360 (for instance, on James Bruton's nearly 12' tall, world record sculpture of himself, seen on page 118). We the Builders utilized Netfabb to manually dissect their projects. This will all take some skill, knowledge, and effort.

MESHMIXER (meshmixer.com)
Good ol' Meshmixer has been around for years and is still a highly recommended tool for simple modifications and model mash-ups. The cutting tool in this package is more powerful than the one found in most slicers and can allow you to get more fine control over chopping up a model. There will still be some manual effort as you'll need to determine every single slice placement and create your own alignment structures (Figure **D**).

LUBAN (luban3d.com)
The best tool for this currently is Luban. Fairly new to the scene, Luban is capable of splitting your model automatically in various ways, while also automatically adding alignment methods, and even numbering the parts if you need.

The package has many options on how it slices, what alignment method you'd like to use, the size of slices, and more. Then, there are methods of manually determining how things are sliced if, for example, you want a split to happen along a complicated contour (Figure **E**).

Attaching Your Parts
After you get all the parts printed, you're going to have to make it all stick together. For the most part, cyanoacrylate (aka super glue) will work wonders. However, if you happen to be using a material such as Nylon or TPU you may have to experiment with other methods such as using a two-part epoxy that remains slightly more flexible. ◐

Billz Sharif's *Joker*, dissected by Luban software (top) and assembled.

Large-scale triceratops print by Luban founder Lujie Chen.

Jacob Ayers designed and printed the Benjamin Harrison house with the help of Luban software.

CUT
and Paste

Make a cardboard attachment guide for your classroom to help inspire great builds

Written by Mike Senese

Cardboard is a wonderful prototyping material. Quickly and easily cut out a design for a project, tape it together, and in minutes you can have a rough physical rendition of your concept to examine and test out. If it looks good, you can move to a more permanent material for the actual build. But if not, no problem — simply redesign the problematic pieces, replace the old ones with new, and see if everything fits and functions as desired. Depending on the build, you can make a cardboard prototype in almost no time at all, and with the availability of cardboard (especially if you shop online as much as I do), it's surprisingly easy to get access to free material.

Because of its simplicity, cardboard is especially useful in classrooms. To inspire young minds with ways to utilize the material in their builds, some classes are implementing their own "cardboard attachments" posters, showing various ways to affix different structures together. We've traced the concept back to Festus Elementary art teacher Sarah Wyman, while the version shown here was made by the Exploratorium's Deanna Gelosi for a workshop in their Tinkering Studio. Many others exist with expanded elements; if you make one too, we'd love to see it! ●

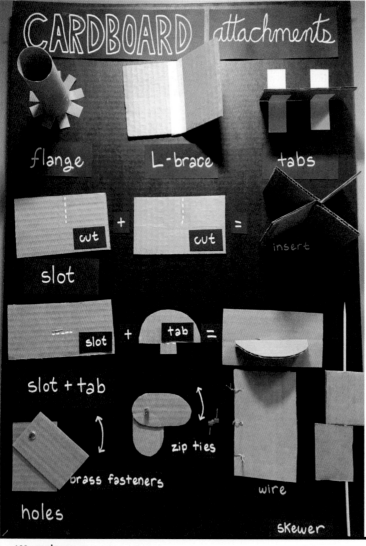

CARDBOARD attachments

flange

L-brace

tabs

cut + cut = insert

slot

slot + tab =

slot + tab

zip ties

brass fasteners

wire

holes

skewer

Ryoko Matsumoto

Maker Faire Shenzhen

Maker Faire®

LEATHERMAN FREE P2

$119 leatherman.com

Leatherman's new "free" series has a cure for something that has long been a pain point for me. The Free P2 is fully usable with only one hand. You can completely open and close every aspect of this 19-tool system without needing to employ extra fingers, and it isn't awkward or difficult. The thumb-press lifting mechanism alone has removed all frustration from opening the tools you want.

The construction feels solid, there's no rattle if you shake the tool, and everything snaps into place with a satisfying click. As someone who abuses their tools a bit, I appreciate the replaceable blades on the wire cutter.

If I really had to find something to complain about, I'd say that the lock spring gets in the way of using the fancy thumb-press release on the biggest blade, but the blade is shaped in a way that you can engage it one handed without this new feature. –*Caleb Kraft*

ADAFRUIT NEOTRELLIS M4

$60 (full kit) OR $40 (board only)
adafruit.com

The NeoTrellis has a nifty microcontroller at its heart, but the real draw is what it powers: A reprogrammable light-up keypad. Adafruit's always-excellent tutorials will get you to load up a drum machine, or put sound effects at your fingertips, one per button.

The NeoTrellis has a headset jack and microphone input included. Aside from that, there are just a few pins accessible to the outside world. If you wanted to attach a load of sensors and actuators, Adafruit has no shortage of other better-suited boards.

This one is a pre-built I/O and audio device that lets you use those microcontroller smarts to take over how it performs. You have a field of 24 glowing buttons to work with. Each button can change color as you see fit. Outputting sound in response to button presses is the obvious use.

I hope we'll see more devices like the NeoTrellis kit: Ones that give the user a head start on the hardware side of their creation.

–Sam Brown

ENGRAVING BALL VISE

$129 amazon.com/dp/B00BR4XWRG

Ball vises are common in engraving, jewelry arts, and other crafts that require fine rotation in a vise. They can be pretty expensive, too. This is an economy version of some of the fancier ones available, costing a fraction of their prices. It may not be quite as nice, but for the savings it is hard to complain.

This set comes with a bunch of different workholding options. The fit can be a bit sloppy sometimes, which can lead to difficulty when clamping finicky work, but that hasn't been too much of an issue for me so far. After a bit of additional lubrication, it spins freely and the adjustable brake is nice, so you can finely tune the resistance.

For a beginner, this is a great cost-saving option that will get you working quickly.

–Caleb Kraft

NVIDIA JETSON NANO DEVELOPER KIT

$99 nvidia.com

It's a good time to be a maker in artificial intelligence. Nvidia's Jetson Nano is the fourth board in their Jetson line for running robots and other autonomous machines. But now, at just $99, Jetson is affordable to a hobby market. If you want to try out deep learning to make a self-driven robot or other trained-rather-than-programmed creation, Jetson Nano occupies a sweet spot of power and price.

Nano is built for flexibility. It runs all the popular machine learning frameworks. If drawing 10W of power (comparable to a Raspberry Pi) is too much, it can down-power half its cores to run at 5W. And if you discover that you just can't do without faster performance, the same deep learning models you built on Nano will run on the higher-end Jetsons, no changes needed.

Nvidia has launched Jetson with a few first projects to try out: HelloAI introduces Machine Learning concepts. Makers may want to skip straight to JetBot, a robot that can be put together for $250 in parts, Jetson included. Not only can it dodge objects, it'll be able to identify those objects, too.

–Sam Brown

TEMPLATE MAKER

FREE templatemaker.nl

Sometimes a resource pops up that is just over-the-top in terms of its usefulness. Template Maker by Maarten van der Velde is a collection of shape calculators that create various papercraft templates, all free. At first glance, you might expect simple things like boxes and cylinders of various sizes, but as you dig deeper you'll find the shape calculators are not only varied, but the level of customization of each is impressive.

Each calculator offers the ability to modify the shape, output a template immediately, then save it to your computer for later use. You can quickly switch between units of measurement like inches, centimeters, and millimeters, and change aspects of the shape itself, like the number of points on a star. There are mostly bold shapes that would be convenient packaging for other things, but there are also interesting structures meant to hold things inside a box. I could easily see returning to this page in the future to help projects along. *–Caleb Kraft*

SHOW & TELL

Get inspired by some of our favorite submissions to the Make: community

① UPCYCLED HYDROPONIC PLANTER
Sandy Roberts set up this easy hydroponics planter from a used plastic water bottle and an old cotton sock. It's a clever and inexpensive way to get started growing hydroponic plants.
makershare.com/projects/upcycled-hydroponics-planter

② WIZARD STAFF
Steve Wyness masterfully built this unique light-up wizard's staff out of a tree root, real quartz crystals, and some electronics components. The organic shape of the wood drove much of the design, including the puzzle-worthy placement of the electronics and crystals. Wyness has some great tips if you decide to design your own.
makershare.com/projects/wizard-staff-colored-leds-and-flash-effect

③ LET THE SUN SHINE
Sometimes it pays off to experiment without a set goal in mind. While **Dave Dunlop** was toying with a new light sensor, his tinkering led to this Raspberry Pi-based build that automatically opens or closes the blinds based on the light coming in through the window.
makershare.com/projects/let-sun-shine

④ SOCIAL MEDIA WITHOUT THE INTERNET
"What would it be like to interact with people using 'social media features' in real life?" That's the question artist and designer **Tuang T** tries to answer with this sensor-filled jacket. A physical handshake increase your friend count and a high five adds a "like" to the running count on a large digital display. Watch the video for a full run-down.
makershare.com/projects/social-media-without-internet

Sandy Roberts, Steve Wyness, Dave Dunlop, Tuang T